GALILEO.

THE

ORBS OF HEAVEN,

OR,

THE PLANETARY AND STELLAR WORLDS.

LORD ROSSE'S TELESCOPE, PARSONSTOWN.

LONDON:

OFFICE OF THE NATIONAL ILLUSTRATED LIBRARY.

198, STRAND.

THE

ORBS OF HEAVEN,

OR,

THE PLANETARY AND STELLAR WORLDS,

A POPULAR EXPOSITION OF

THE GREAT DISCOVERIES AND THEORIES OF
MODERN ASTRONOMY

BY O. M. MITCHELL, A.M.

Director of the Cincinnati Observatory.

LONDON:

OFFICE OF THE NATIONAL ILLUSTRATED LIBRARY,

198, STRAND.

1851.

LONDON:
BRADBURY AND EVANS, PRINTERS, WHITEFRIARS.

PREFACE.

THE following volume, by an American Author, contains a graphic and popular exposition of the great discoveries of astronomical science; and originated, as the Author informs us, under the following circumstances. The writer some time ago conceived the idea of erecting a great astronomical observatory in the city of Cincinnati. His attention had been for many years directed to this subject, by the duties of the professorship which he held in the College. "In attempting to communicate the great truths of Astronomy, there were no instruments at hand, to confirm and fix the facts recorded in the books. Up to that period America, and the western portions particularly, had given but little attention to practical Astronomy. An isolated telescope was found here and there scattered through the country; but no regularly organised observatory, with powerful instruments, existed within the limits of the United States, so far as the writer knew."

To ascertain whether any interest could be excited in the public mind, in favour of Astronomy, a series of lectures was delivered in the Hall of the Cincinnati College. Encouraged by the large audiences which continued through two months to fill the lecture room, and still more by the request to repeat the last lecture of the course in one of the great churches of the city, the Author matured a plan for the building of an observatory. In pursuance of that plan, and in consequence of the favour with which it was received, he visited Europe, procured instruments, examined observatories, and obtained the requisite knowledge to erect and conduct the institution which it was hoped would be one day reared.

The hope was not groundless ;—subscriptions poured in, though not without difficulty ;—a large refractive telescope was purchased, and a piece of ground was generously given by a private individual for the site of an observatory. The work was pushed rapidly forward. In February, 1845, the great telescope safely reached the city of Cincinnati ; and in March the building was ready for its reception.

The ten lectures which form this volume, were those to which the observatory owes its origin. The object of the lectures is to trace the career of the human mind, in its researches among the stars. "No one science," says the Author, " so perfectly illustrates the gradual growth and development of the powers of human genius as Astronomy. The movement of the mind has been constantly onward ; its highest energies have ever been called into requisition ;

and there never has been a time when Astronomy did not
present problems not only equal to all that man could do,
but passing beyond the limits of his greatest intellectual
vigour.   Hence, in all ages and countries, the absolute
strength of human genius may be measured by its reach to
unfold the mysteries of the stars.

" It will be seen that in the following lectures one single
object has engaged the attention of the writer : *the structure
of the universe, so far as revealed by the mind of man.*

" The uses of science have in no way been considered. The
effects on the mind, on society, on civilisation, on commerce,
on religion, have not been permitted to mar the unity of
the original design.   The onward, steady, triumphant march
of mind, in its study and exploration of the universe of God,
has been the only object, the single theme of the entire
series."

CONTENTS.

LECTURE VIII.

LECTURE IX.

LECTURE X.

APPENDIX.

BY PROFESSOR OLMSTED, LL.D.

LIST OF ILLUSTRATIONS.

THE

ORBS OF HEAVEN.

INTRODUCTORY.

AN EXPOSITION OF THE PROBLEM WHICH THE HEAVENS PRESENT FOR SOLUTION.

THE subject to which your attention is invited, claims no specific connection with the every day struggle of human life. Far away from the earth on which we dwell, in the blue ocean of space, thousands of bright orbs, in clusterings and configurations of exceeding beauty, invite the upward gaze of man, and tempt him to the examination of the wonderful sphere by which he is surrounded. The starry heavens do not display their glittering constellations in the glare of day, while the rush and turmoil of business incapacitate man for the enjoyment of their solemn grandeur. It is in the stillness of the midnight hour, when all nature is hushed in repose, when the hum of the world's on-going is no longer heard, that the planets roll and shine, and the bright stars, drooping through the deep heavens, speak to the willing spirit that would learn their mysterious being.

B

Often have I swept backward in imagination six thousand years, and stood beside our Great Ancestor, as he gazed for the first time upon the going down of the sun. What strange sensations must have swept through his bewildered mind, as he watched the last departing ray of the sinking orb, unconscious

Adam contemplating the setting sun.

whether he should ever behold its return. Wrapt in a maze of thought, strange and startling, his eye long lingers about the point at which the sun had slowly faded from his view. A mysterious darkness, hitherto unexperienced, creeps over the face of nature. The beautiful scenes of earth, which, through the swift hours of the first wonderful day of his existence, had so charmed his senses, are slowly fading one by one from his dimmed vision. A gloom, deeper than that which covers earth, steals across the mind of earth's solitary inhabitant. He raises his inquiring gaze towards heaven, and lo! a silver crescent of light, clear and beautiful, hanging in the western sky, meets his astonished eye. The young moon charms his untutored vision, and leads him upward to her bright attendants, which are now stealing, one by one, from out the deep blue sky. The solitary gazer bows, and wonders, and adores. The hours glide by; the silver moon is gone; the stars are rising, slowly ascending the heights of heaven, and solemnly sweeping downward in the stillness of the night. The first grand revolution to mortal vision is nearly completed. A faint streak of rosy light is seen in the east; it brightens: the stars fade; the planets are extinguished: the eye is fixed in mute astonishment on the growing splendour, till the first rays of the returning sun dart their radiance on the young earth and its solitary inhabitant. To him "the evening and the morning were the first day."

The curiosity excited on this first solemn night, the consciousness that in the heavens God had declared his glory, the eager desire to comprehend the mysteries that dwell in these bright orbs, have clung to the descendants of him who first watched and wondered, through the long lapse of six thousand years. In this boundless field of investigation, human genius has won its most signal victories. Generation after generation has rolled away, age after age has swept silently by, but each has swelled by its contribution the stream of discovery. One barrier after another has given way to the force of intellect; mysterious movements have been unravelled; mighty laws have been revealed; ponderous orbs have been weighed, their reciprocal influences computed, their complex wanderings made clear; until the mind, majestic in its strength, has mounted step by step up the rocky height of its self-built pyramid, from whose star-crowned summit it looks out upon the grandeur of the universe, self-clothed with the prescience of a God. With resistless energy it rolls back the tide of time, and lives in the configuration of rolling worlds a thousand years ago; or, more wonderful, it sweeps away the dark curtain from the future, and beholds those celestial scenes which shall greet the vision of generations when a thousand years shall have rolled away, breaking their noiseless waves on the dim shores of eternity.

To trace the efforts of the human mind in this long and ardent struggle; to reveal its hopes and fears, its long years of patient watching, its moments of despair and hours of triumph; to develope the means by which the deep foundations of the rock-built pyramid of science have been laid, and to follow it as it slowly rears its stately form from age to age, until its vertex pierces the very heavens; these are the objects proposed for accomplishment, and these are the topics to which I would invite your earnest attention. The task is one of no ordinary difficulty. It is no feast of fancy, with music and poetry, with eloquence and art, to enchain the mind. Music is here; but it is the deep and solemn harmony of the spheres. Poetry is here; but it must be read in the characters of light, written on the sable garments of night. Architecture is here; but it is the colossal structure of sun and system, of cluster and universe. Eloquence is here; but "there is neither speech nor language: its voice is not heard." Yet its resistless sweep comes over us in the mighty periods of revolving worlds.

Shall we not listen to this music, because it is deep and solemn? Shall we not read this poetry, because its letters are the stars of heaven? Shall we refuse to contemplate this architecture, because "its architraves, its archways, seem ghostly from infinitude?" Shall we turn away from this surging eloquence, because its utterance is made through sweeping worlds? No: the mind is ever inquisitive, ever ready to attempt to scale the most rugged steeps. Wake up its enthusiasm, fling the light of hope on its pathway, and no matter how rough, and steep, and rocky it may prove, ONWARD! is the word which charms its willing powers.

It is not my wish or design to introduce you to the dark technicalities of science, neither do I propose to rest satisfied with the barren statement of the results which have been reached by the efforts of genius. While on the one hand I shall endeavour to shun all attempt at critical scientific demonstration, which could only be intelligible to the professed student of astronomy, I shall on the other hand fearlessly attempt such an exposition of the processes and trains of reasoning by which great truths have been elicited, as to show to every intelligent mind that the problem is not impossible; by simplicity of language, by familiar illustrations, to fling light enough upon these mysterious propositions;—to show a pathway, though it be dim and rugged, still a pathway, which, if pursued, shall certainly lead to a full and perfect solution. I ask, then, no critical previous knowledge of the subject, on the part of those who would follow me in the wonderful developments which I am about to attempt. Give me but your earnest and unbroken attention. Go with me in imagination, and join in the nightly vigils of the astronomer, and while his mind with powerful energy struggles with difficulty, join your own sympathetic efforts with his, hope with his hope, tremble with his fears, and rejoice with his triumphs. Lend me but this kind of interest, and my task is already half accomplished.

Before proceeding to an actual exposition of the structure of the Heavens, I propose in this introductory lecture to announce the nature of the problem, which the mind has essayed to resolve, and to point out the more important auxiliaries, mental and mechanical, which it has conjured to its aid. If the difficulties of this problem should overwhelm the mind, let it be remembered that the astronomer has ever lived, and never dies. The sentinel upon the watch-tower is relieved from duty; but another takes his place, and the vigil is unbroken. No: the astronomer never dies. He commences his investigations on the hill tops of Eden; he studies the stars through the long centuries of antediluvian life. The deluge sweeps from the earth its inhabitants, their cities and their monuments; but when the storm is hushed, and the heavens shine forth in beauty, from the summit of Mount Ararat the astronomer resumes his endless vigils. In Babylon he keeps his watch, and among the Egyptian priests he inspires a thirst for the sacred mysteries of the stars. The plains of Shinar, the temples of India, the pyramids of Egypt, are equally his watching-places. When science fled to Greece, his home was in the schools of her philosophers; and when darkness covered the earth for a thousand years, he pursued his never-ending task from amidst the burning deserts of Arabia. When science dawned on Europe, the Astronomer was there, toiling with Copernicus, watching with Tycho, suffering with Galileo, triumphing with Kepler. Six thousand years have rolled away since the grand investigation commenced. We stand at the terminus of this vast period, and, looking back through the long vista of departed years, mark with honest pride the successive triumphs of our race. Midway between the past and future, we sweep backward and witness the first rude effort to explain the celestial phenomena: we may equally stretch forward thousands of years, and although we cannot comprehend what shall be the condition of astronomical science at that remote period, of one thing we are certain; the past, the present, and the future, constitute but one unbroken chain of observations, condensing all time, to the astronomer, into one mighty *now*.

From the vantage ground which we occupy, it will not be difficult to announce so much of the great problem as has already been resolved, and to form some approximate conception of what remains for future ages to accomplish.

In the exposition about to be attempted, I do not propose to present any trains of reasoning, or any results which may have been reached. These shall engage our attention hereafter. At present, permit me simply to translate into language the questions which the visible heavens propound.

The most cursory examination of the celestial vault reveals the fact, that not one solitary object, visible to the eye, is at rest. Motion is the attribute of sun and moon, and planets and stars. The earth we inhabit alone remains fixed, to the senses.

The first great problem propounded for human ingenuity, is to sever *real motion* from that which is *unreal* and only *apparent*. To accomplish this, some knowledge of the form of the earth which we inhabit must be obtained. Not only must we acquire a knowledge of its figure, but in like manner we must learn with certainty its actual condition, whether of rest or motion. If at absolute rest in the centre of the universe, then the rising sun, the setting moon, the revolving heavens, are real exhibitions, and must be examined as such. On the contrary, should it be found to be impossible to predicate of the earth absolute immobility, then arises the complicated question, how many motions belong to it? and with what velocity does it move? If a motion of rotation exist, what is the position of the axis about which it revolves, and is this axis permanent or changeable? If a motion of translation in space must be adopted, then whither is the earth urging its flight? what the nature of the path described? the velocity of its movement, and the laws by which it is governed? These are some of the questions which present themselves in the outset, touching the condition of the earth, on whose surface the astronomer is located, in his researches over the heavens.

Beyond the limits of the earth, a multitude of objects present themselves for examination : and first of all the sun, the great source of life, and light, and heat, demands the attention of the student of the heavens. That some inscrutable tie binds it to the earth, or the earth to it, was early recognised in the fact, that whether the sun was moving or at rest, the relative distance of it and the earth never changed by any great amount ; and whatever changes did occur, were all obliterated in a short period, and the distance by which these bodies are separated was restored to its primitive value, to recommence its cycle of changes in the same precise order. Here, then, was a grand problem : to determine the relations existing between the sun and earth ; to endue with motion that one of these bodies which did move ; and to fix the limits within which the observed changes occurred, both in time and distance.

While the connection between the sun and earth was certain, a mutual dependence between the earth and the other great source of light, the moon, was equally manifest. The invariability in the apparent diameter of the moon demonstrates the fact, that whether the earth were moving or stationary, the moon never parts company with our planet. In all her wanderings among the fixed stars, in her elongations from the sun, in her wondrous phases and perpetual changes, some invisible hand held her at the same absolute distance from the earth. But to decide whether this power resided in the earth or the moon, or in both ; to explain these wondrous changes from the silver crescent of the western sky to the full orb which rose with the setting sun, pouring a flood of light over all the earth ; to develop the mysterious connection between the disappearance of the moon and those terrific phenomena, the going out of the sun in dim eclipse ; these furnished themes for investigation, requiring long centuries of patient watching, and of never-ending toil.

Passing out from the sun and moon to the more distant stars, among the brightest of those which gemmed the nocturnal heavens, a few were found differing from all the rest in the fact that they wandered from point to point, and, at the end of intervals widely differing among themselves, swept round the entire heavens, and returned to their starting point, to recommence their ceaseless journeys. These were named planets, *wanderers*, in contradistinction to the host of stars which were fixed in position, unchanged from century to century.

Hence arose a new and profound series of investigations :

Where were these wandering stars urging their flight? Were their motions real or apparent? Were their distances equal or unequal? Did any tie bind them to the earth, or to the sun, or to each other? Were their distances from the earth constant or variable? Were their motions irregular, or guided by law? Did they accomplish their revolutions among the fixed stars in regular curves, or in lawless wanderings? Among all the moving bodies, sun, moon, and planets, could any principle of association be traced which might bind them together, and form them into a common system?

To resolve these profound questions, a critical watch is kept on all the moving bodies. Their pathway is among the stars, and to these ever-during points of light their positions are constantly referred. If beyond the limits of the moving bodies a dark veil had been drawn so as to have excluded the light of the stars, at the first glance it might seem that by such a change simplicity would have been introduced, and the perplexity arising from the motion of the planets among the profusely scattered stars would have been removed. But let us not judge too hastily. Blot out the stars, and give to the sun, moon, and planets, a blank heavens in which to move, and the possibility of unravelling their mysterious motions, mutual relations, and common laws is gone for ever.

This will become manifest when we reflect that on such a change, not a fixed point in all the heavens would remain, to which we could refer a moving planet. They must then be referred to each other, and the motion due to the one, would become inextricably involved in that due to the other, and neither could be determined with any precision. Like the ocean islands which guided the early mariners, so God has given to us the stars of heaven as the fixed points to which we can ever refer, in all parts of their revolution, the places of the wandering planets, and the swiftly revolving moon.

As the necessity for accuracy in watching the movements of the planets became more apparent, the attention was directed to the acquisition of the means by which this might be accomplished. Hence we find in the earliest ages the astronomer grouping the fixed stars into constellations, breaking up the great sphere of the heavens into fragments, the more easily to study its parts in detail. Not only are the stars of each constellation numbered, their brilliancy noted, but their relative places in the constellation and to each other are fixed with all the precision which the rude means then in use permitted. Names are fixed to these different groupings; when, or where, or by whom, we know not. Neither history nor tradition leads us back to this first breaking up of the heavens, but the names then bestowed on the fragmentary parts, the richer constellations, have survived the fall of empires, and are fixed for ever in the heavens.

Chaldean Shepherds naming the Constellations.

Possessing now a thorough knowledge of the objects among which the planets were moving, and the means of measuring with approximate accuracy, their distance from the stars along their path, it became possible to trace a planet in its career, and to note the changes of its velocity.

New and wonderful discoveries were thus made. It was found
that all the planets moved with an irregular velocity. Sometimes
swiftly advancing among the fixed stars, then slowly relaxing
their speed, they actually stopped, turned backward in their
career, stopped again, and then, at first slowly, but afterwards
more rapidly, resumed their onward motion. These strange
and anomalous motions, differing from anything remarked in the
sun and moon, furnished new themes for discussion, new problems
for solution. While the phenomena above alluded to became
known, the same chain of observations revealed the remarkable
fact, that the periods of revolution of the planets, though differing
for each one of the group, were identical for any one individual ;
and moreover, that a simple curve marked out the pathway of
sun, moon, and planets, among the fixed stars, and that all
these wandering bodies were confined to a narrow zone or belt
in the heavens.

Centuries had now rolled away, nay, even thousands of years
had slowly glided by, since the mind had first given itself to the
examination of the heavens ; and while discovery after discovery
had rewarded the zeal of the observer in every age, yet the
grand object of research, the distinction between actual and
apparent motion, had thus far eluded the utmost efforts of
human genius. But a brighter day was dawning. Each successive effort tore away some petty obstruction which impeded
the march of mind upward towards the lofty region of truth.
Facts grew and multiplied. Phenomena, striking and diversified,
were collated and compared. The mind in imagination took
leave of the earth as the centre of all these complex movements,
inexplicable on its surface, and naturally urged its flight towards
the sun. There it paused and rested, and from this fixed point
looked out upon the circling orbs, and lo ! the complexity of
their movements melted away. The centre was found, the
mystery solved, the ponderous earth rescued from its false position, rolled in its place among the planets, one of the great family
that swept in beauty and harmony about their common parent
the sun.

The mind now stood upon the first platform of the rocky
pyramid which it had been slowly rearing, and with which it
had been slowly rising, through long centuries of ceaseless toil.
One grand point had been gained. Darkness had given way to
light, but the great problem of the universe was yet to be

resolved. All this long and arduous struggle had only revealed what the problem was. Appearances were now separated from realities, and with a fresh and invigorated courage the human mind now gave its energies to the accomplishment of definite objects, no longer working uncertainly in the dark, but with the clear light of truth to guide and conduct the investigation.

Possessed of these extraordinary advantages, the advance now became rapid and brilliant, as it had previously been slow and discouraging. That the planets, reckoning the earth as one, constituted a mighty family of worlds, was now manifest; whether linked singly to the sun, or mutually influencing each other, was the grand question. This great problem rested upon the resolution of a multitude of subordinate ones. The actual curve constituting the planetary orbits, the magnitude of these orbits, their actual position in space, the values and directions of their principal lines, the laws of their motion, all these and many more questions of equal importance and intricacy presented themselves in the outset of the examination now fairly commenced. Human skill was exhausted in the contrivance and construction of mechanical aids by which the movements of the planets might be watched with the greater accuracy. Partial success crowned these extraordinary efforts, but there yet remained delicate investigations which with the utmost skill in observing escaped the farthest reach of man's eagle gaze, and seemed to bid defiance to all his powers.

To conquer these difficulties, one of two things must be accomplished; either man must sweep out from earth towards the distant planets, to gain a nearer and more accurate view, or else bring them down from their lofty spheres to subject themselves to his scrutinising gaze. How hopeless the accomplishment of either of these impossible alternatives! But who shall prescribe the limits of human genius? In studying the phenomena of the passage of light through transparent crystallised bodies, a principle was discovered which let in a gleam of hope on the disheartened mind. It seized this principle, converted it to its use, and armed itself with an instrument more wonderful than any that fancy in its wildest dreams ever pictured to the imagination. With the potent aid of this magic instrument, the astronomer was no longer bound hopelessly to his native earth; without indeed quitting in person its surface, his eye, gifted with superhuman power, ranged the illimitable fields of space. He visited the moon,

and found a world with its lofty mountains and spreading valleys.
The star-like planets swelled into central worlds, with their
circling moons ; and myriads of fixed stars, hitherto beyond the
reach of human vision, stood revealed in all their sparkling
beauty. It was as if the united ranges of a thousand eyes were
all concentrated in a single one.

A new era now dawns on the world. The delicate and invisible
irregularities of the planetary motions are now fully revealed,
and the data rapidly accumulate by means of which the last grand
question is to be resolved. The orbital curves are determined.
The laws of the revolving planets are revealed. A mysterious
relation between the distances of the planets from the sun and
their periods of revolution, unites them positively into one grand
family group. That they are bound to the sun by some inscrutable power, is certain, and it now remains to determine the law
of increase and decrease of this force for all possible distances.
This last truth is finally achieved, and the wisdom of God is
vindicated in the beautiful structure of our grand system.

The second lofty platform is reached in the mighty pyramid,
whose summit is now nearing the stars of heaven. From this
elevation the mind looks out upon the circling planets and their
revolving satellites, and the mysterious comet, and ventures to
propound the question, Do these bodies so interfere with the
movements of each other, as to affect permanently the structure
by which the equilibrium and stability of the entire system is
guaranteed ?

To answer this question, a new train of investigation is commenced, satellite is weighed against planet, and planet against
the sun, until the mass of matter contained in each individual of
the system becomes accurately known.   Then is undertaken the
grand problem of perturbations. The telescope reveals the fact
that slow and mysterious changes are going on in the mean
motions of the moon, in the figure of the planetary orbits, and
in the relative positions which these orbits hold to each other.
Are these changes ever progressive ? If this be true, then does
the system contain within itself the seeds of decay, the elements
of its own destruction. Slowly but surely as the solemn tread
of time, the end must come, and, one by one, planet and satellite
and comet, must sink for ever in the sun ! Long and arduous
was the struggle to reach the true answer to this difficult question.
The entire solution involved a multitude of parts.

When the mutual dependence of the multitude of bodies constituting our system was discovered, when planet, and satellite, and comet, were found to feel and sway to the influence which each exerted on the other, the simplicity of their movements was gone for ever; orbits once fixed in the heavens, slowly swung away from their moorings; the beautiful precision which had to all appearance marked the planetary curves, was destroyed. The regularity of their motions was changed into irregularity, and a system of complexity which seemed to bid defiance to all effort at comprehension, presented itself to the human intellect.

It was no less than this : given, a system of revolving worlds, mutually operating on each other; required, their magnitudes, masses, distances, motions, and positions, at the close of a thousand revolutions. What mind possessed the gigantic power to grasp this mighty problem ? Reason was lost in wandering mazes, and the brightest intellect sunk clouded in gloom.

In this dilemma, the mind turns inward on its own resources. As the physical man climbs some mountain height by successive efforts, rising higher and higher, scaling rock after rock, and mounting precipice after precipice, by the use of strength comparatively feeble, resting and recruiting as it becomes exhausted; was it impossible to contrive some mental machinery which might give to the reason the power of prosecuting its difficult researches, in such a manner that it might stop and rest, and not lose what it had already gained in its onward movement ?

Geometry had invigorated the reason, as exercise toughens and strengthens the muscles of the human frame. But it had given to the mind no mechanical power, wherewith to conquer the difficulties which rose superior to its natural strength. Archimedes wanted but a place whereon to stand, and with his potent lever he would lift the world. The astronomer demands an analogous mental machinery to trace out the complex wanderings of a system of worlds. What the human mind demands and resolves to find, it never fails to discover. The infinitesimal analysis is reached, its principles developed, its resistless power compelled into the service of human reason. I shall not now stop to explain the nature of this analysis. Its power and capacity alone engage our attention at the present. Once having seized on a wandering planet, it never relaxes its hold; no matter how complicated its movements, how various the

influences to which it may be subjected, how numerous its revolutions, no escape is possible. This subtle analysis clings to its
object, tracing its path and fixing its place with equal ease,
at the beginning, middle, or close of a thousand revolutions,
though each of these should require a century for its accomplishment.

Armed with this analysis, which the mind had created for its
use, giving it a strength only commensurate with the increased
power which had been given to the human eye, it concentrated
its energies once more upon this last greatest problem. One by
one these strong-holds give way, the resistless power of analysis
marches onward from victory to victory, until finally the sublime
result is reached, the system is stable, the equilibrium is perfect ;
slowly rocking to and fro in periods which stun the imagination,
the limits are prescribed beyond which these fluctuations shall
never pass.

Here it would seem that human ambition might rest. Satisfied
with having mastered the mysteries of the system with which
we are united, the mind might cease its arduous struggle, and
leave the wilderness of fixed stars free from its intrusions and
ceaseless persecutions. But this is not the effect produced by
victory ; success but engenders new desires, and prompts to
more difficult enterprises. Man having obtained the mastery
over his own system, boldly wings his flight to the star-lit vault,
and resolves to number its countless millions, to circumscribe its
limitless extent, to fathom its infinite depth, to fix the centre
about which this innumerable host is wheeling its silent and
mysterious round.

Here commences a new era. The first step in the stupendous
enterprise is to determine the distance of some one fixed star.
Here again the mind is long left to struggle with difficulties
which it seemed that no ingenuity or skill could remove. But
its efforts do not go unrewarded. If it fails in the accomplishment of its grand object, it is rewarded by the most brilliant
discoveries. The mighty law governing the planetary worlds is
extended to the regions of the fixed stars ; motion is there
detected, orbital motion, the revolution of sun about sun. The
swift velocity of light is measured, to become the future unit in
the expression of the mighty distances which remain yet to be
revealed. Ever baffled but never conquered, the mind returns
again and again to the attack, till finally the problem slowly

yields, the immeasureable gulf is passed, and the distance of a single star rewards the toils of half a century. But what a triumph is this! it is no less than a revelation of the scale on which the universe is built. The interval from sun to fixed stars, is that by which the stars are separated; and a reach of distance is opened up to the mind, which it only learns to contemplate by long continued effort.

But another startling fact is revealed in the prosecution of these profound investigations. The minute examinations of the fixed stars have changed their character. For thousands of years they have been regarded as absolutely fixed among each other. This proves to be mere illusion, resulting from the use of means inadequate for the determination of their minute changes. Under the scrutinising gaze of the eye, with its power increased a thousand-fold, the millions of shining orbs which fill the heavens, are all found to be slowly moving around each other, slowly, as seen from our remote position, but with amazing velocity when examined near at hand.

A new problem of surprising grandeur now presents itself. Are these motions real? or are they due to a motion in the great centre of our system? A series of examinations analogous to those which divided between the real and apparent motions in the planets, is commenced and prosecuted with a zeal and devotion unsurpassed in the history of science. The mind rises to meet the sublime investigation. For a hundred years it toils on; again it triumphs, the truth is revealed. The immobility of the sun is gone for ever; our last fixed point is swept from under us, and now the entire universe is in motion.

With redoubled energy the mind still prosecutes the inquiry, whither is the sun sweeping, and with what velocity does it pursue its unknown path? Strange and incredible as it may appear, these questions are answered; and in attaining this answer, the means are reached to separate between the real and apparent motions of the fixed stars, and to study their complex changes, and to rise, by slow degrees, to a complete knowledge of the movements of the grand sidereal system. Here we pause. Rapidly have we descended the current of astronomical research, we have attained the boundary of the known. We stand on the dim confines of the unknown. All behind us is clear, and bright, and perfect; all before us is shrouded in gloom, and darkness, and doubt. Yet the twilight of the known flings its feeble light into the domain of the unknown ; and we are permitted to gather some idea, not of all that remains to be done, but of that which must be first accomplished.

Let us then stretch forward and propound some of those questions which nature yet presents for solution, but which have hitherto resisted the efforts of the human mind. First of all, we begin with our own system. How came it to be constituted as it actually exists ? All the analogies of nature forbid the idea that it was thus instantly called into being by the fiat of Omnipotence. Does it come, then, from some primitive modification of matter, under the action of laws working out their results in countless millions of ages ? Who shall present the true cosmogony of the solar system ?

But this is only one unit among many millions. Whence the myriads of stars ? those stupendous aggregations into mighty clusters ? what the laws of their wonderful movements, of their perpetual stability ? Who will explain the periodical stars, that wax and wane, like the changing moon; or, still more wonderful, reveal the mystery of those which have suddenly burst on the astonished vision of man, and have as suddenly gone out for ever in utter darkness ?

Such are the questions which remain for the resolution of future ages. We may not live to witness these anticipated triumphs of mind over matter ; but who can doubt the final result ? Look backward to the Chaldean shepherd, who watched the changing moon from the plains of Shinar, and, wondering, asked if future generations would reveal those mysterious phases. Compare his mind and knowledge with those of the modern astronomer, who grasps, at a single glance, the past, present, and future changes of an entire system. Are the heights which remain to be reached more rugged, more inaccessible, than those which have been already so triumphantly scaled ? The observations recorded in Babylon three thousand years ago have reached down through the long series of centuries, and are of inestimable value in the solution of some of the darkest problems with which the mind has ever grappled. In like manner, the records we are now making shall descend to unborn generations, and contribute to effect the triumphs of genius when three thousand years shall have rolled away. If doubt arises as to the final resolution of these profound questions, from the immense distance of the objects under examination, let us call to mind the fact, that the artificial eye which man has furnished for his use, possesses a glance so piercing, that no distance can hide an object from his searching vision.

Should Sirius, to escape this fiery glance, dart away from its sphere, and wing its flight at a velocity of twelve millions of miles in every minute, for a thousand years ; nay, should it sweep onward at the same speed for ten thousand years, this stupendous distance cannot bury it from the persecuting gaze of man. But if distance is to form no barrier, no terminus to these investigations, surely there is one element which no human ingenuity can overcome. The complex movements of the planetary orbs have been revealed, because they have been repeated a thousand times under the eye of man, and from a comparison of many revolutions, the truth has been evolved. But tens of thousands of years must roll away before the most swiftly moving of all the fixed stars shall complete even a small fragment of its mighty orbit. With motions thus shrouded, these would seem to be in entire security from the inquisitive research of a being whose whole sweep of existence is but a moment, when compared with these vast periods. But let us not judge too hastily. The same piercing vision that follows the retreating star to depths of space almost infinite, is armed with a power so great, that if this same star should commence to revolve around some grand centre, and move so slowly that five millions of years must roll away before it can complete one circuit, not even a single year shall pass before its motion shall be detected ; in ten years its velocity shall be revealed ; and in the lifetime of a single observer its mighty period shall become known.

If human genius is not to be baffled either by distance or time, numbers shall overwhelm it, and the stars shall find their safety in their innumerable millions. This retreat may even fail. The watch-towers of science now cover the whole earth, and the sentinels never sleep. No star, or cluster, or constellation can ever set. It escapes the scrutinising gaze of one astronomer to meet the equally piercing glance of another. East and west, and north and south, from the watch-towers of the four quarters of the globe, peals the solemn mandate, Onward !

Here we pause. We have closed the enunciation of the great problem whose discussion and solution lie before us, a problem whose solution has been in progress six thousand years ; one

c

which has furnished to man the opportunities of his loftiest triumphs, one which has taxed in every age the most vigorous

Pyramids on a starry night.

efforts of human genius; a problem whose successive developments have demonstrated the immortality of mind, and whose sublime results have vindicated the wisdom and have declared the glory of God. You have listened to the enunciation, we now invite you to follow us in the demonstration. And may that Almighty power, which built the heavens, give to me wisdom to reveal, and to you power to grasp, the truths and doctrines wrested by mind from nature in its long struggle of sixty centuries of toil!

Statue of Newton at Cambridge.

LECTURE II.

THE DISCOVERIES OF THE PRIMITIVE AGES.

O those who have given but little attention to the science of astronomy, its truths, its predictions, its revelations, are astonishing ; and but for their rigorous verification would be absolutely incredible. When we look out upon the multitude of stars which adorn the nocturnal heavens, scattered in bright profusion in all directions, apparently without law, and regardless of order ; when, with telescopic aid, thousands are increased to millions, and suns, and systems, and universes rise in sublime perspective, as the visual ray sweeps outward to distances which defy the powers of arithmetic to express ; how utterly futile does it seem for the mind to dare to pierce and penetrate, to number, weigh, measure, and circumscribe these innumerable millions ! It is only when we remember, that from the very cradle of our race, strong and powerful minds have, in rapid and continuous succession, bent their energies upon the solution of this grand problem, that we can comprehend how it is that light now breaks in upon us from the very confines of the universe, dimly revealing the mysterious forms which lie yet half concealed in the unfathomable gulphs of space. When I reflect on the recent triumphs of human genius ; when I stand on the shore of that mighty stream of discovery, which has grown broader and deeper as successive centuries have rolled away, gathering in strength and intensity, until it has embraced the whole universe of God ; I am carried backward through thousands of years, following this stream, as it contracts towards its source, till finally its silver thread is lost in the clouds and mists of antiquity. I would fain stand at the very source of discovery, and commune with that unknown god-like mind which first conceived the grand thought, that even these mysterious stars might be read, and that the bright page which was nightly unfolded to the vision of man needed no interpreter of its solemn beauties but human genius. There is, to my mind, no finer specimen of moral grandeur than that presented by him who first resolved to read and comprehend the heavens. On some lofty peak he stood, in the stillness of the midnight hour, with the listening stars as witnesses of his vows, and there, conscious of his high destiny, and of that of his race, resolved to commence the work of ages. " Here," he exclaimed, " is my watch-tower, and yonder bright orbs are henceforth my solitary companions. Night after night, year after year, will I watch and wait, ponder and reflect, until some ray shall pierce the deep gloom which now wraps the world."

Thus resolved the unknown founder of the science of the stars. His name and his country are lost for ever. What matters this, since his works, his discoveries, have endured for thousands of years, and will endure as long as the moon shall continue to fill her silver horn and the planets to roll and shine ?

Go with me, then, in imagination, and let us stand beside this primitive observer, at the close of his career of nearly a thousand years, (for we must pass beyond the epoch of the deluge, and seek our first discoveries among those sages whom, for their virtues, God permitted to count their age, not by years, but by centuries,) and here we shall learn the order in which the secrets of the starry world slowly yielded themselves to long and persevering scrutiny. And now let me unfold, in plain and simple language, the train of thought, of reasoning, and research, which marked this primitive era of astronomical science. It is true that history yields no light, and tradition even fails ; but such is the beautiful order in the golden chain of discovery, that the bright links which are known, reveal with certainty those which are buried in the voiceless past. If, then, it were possible to read the records of the founder of astronomy, graven on some

column of granite, dug from the earth, whither it had been
borne by the fury of the deluge, we know now what its hiero-
glyphics would reveal with a certainty scarcely less than that
which would be given by an actual discovery, such as we have
imagined. We are certain that the first discovery ever recorded,
as the result of human observation, was on the *moon*.

The sun, the moon, the stars, had long continued to rise, and
climb the heavens, and slowly sink beneath the western horizon.
The spectacle of day and night was then, as now, familiar to
every eye ; but in gazing there was no observation, and in mute
wonder there was no science. When the solitary observer took
his post, it was to watch the moon. Her extraordinary phases
had long fixed his attention. Whence came these changes ?
The sun was ever round and brilliant, the stars shone with
undimmed splendour, while the moon was ever waxing and
waning—sometimes a silver crescent, hanging in the western
sky, or full orbed, walking in majesty among the stars, and
eclipsing their radiance with her overwhelming splendour.
Scarcely had the second observation been made upon the moon,
when the observer was struck with the wonderful fact, that she
had left her place among the fixed stars, which on the preceding
night he had accurately marked. Astonished, he again fixes
her place by certain bright stars close to her position, and waits
the coming of the following night. His suspicions are confirmed,
—the moon is moving ; and, what to him is far more wonderful,
her motion is precisely *contrary* to the general revolution of the
heavens—from east to west. With a curiosity deeply aroused,
he watches from night to night, to learn whether she will return
upon her track ; but she marches steadily onward among the stars,
until she sweeps the entire circuit of the heavens, and returns
to the point first occupied, to recommence her ceaseless cycles.

An inquiry now arose, whether the changes in the moon, her
increase and decrease, could in any way depend on her place
among the fixed stars. To solve this question, required a longer
period. The group of stars among which the new moon was
first seen was accurately noted, so as to be recognised at the
following new moon, and doubtless our primitive astronomer
hoped to find that in this same group the silver crescent, when it
should next appear, would be found. But in this he was disappointed ; for when the moon became first faintly visible in the
western sky, the group of stars which had ushered her in before,

had disappeared below the horizon, and a new group had taken its place ; and thus it was discovered that each successive new moon fell farther and farther backward among the stars. By counting the days from new moon to new moon, and those which elapsed while the moon was passing round the heavens from a certain fixed star to this same star again, it was found that these two periods were different ; the revolution from new to new occupying 29½ days, while the sidereal revolution, from star to star, required 27⅓ days.

This backward motion of the moon among the stars, must have perplexed the early astronomers ; and, for a long while, it was utterly impossible to decide whether the motion was real or only apparent : analogy would lead to the conclusion that all motion must be in the same direction, and as the heavens revolved from east to west, it seemed impossible that the moon, which manifestly participated in this general movement, should have another and a different motion, from west to east. There was one solution of this mystery, and I have no doubt it was for a long while accepted and believed. It was this : by giving to the moon a slower motion from east to west than the general motion of the heavens, she would appear to lag behind the stars, which would, by their swifter velocity, pass by her, and thus occasion in her the observed apparent motion, from west to east. We shall see presently how this error was detected.

The long and accurate vigils of the moon, and the necessity of recognising her place, by the clusters or groups of stars among which she was nightly found, had already familiarised the eye with those along her track, and even thus early the heavens began to be divided into constellations. The eye was not long in detecting the singular fact, that this stream of constellations, lying along the moon's path, was constantly flowing to the west, and one group after another apparently dropping into the sun, or, at least, becoming invisible in consequence of their proximity to this brilliant orb. A closer examination revealed the fact, that the aspect of the whole heavens was changing from month to month. Constellations which had been conspicuous in the west, and whose brighter stars were the first to appear as the twilight faded, were found to sink lower and lower towards the horizon, till they were no longer seen ; while new groups were constantly appearing in the east.

These wonderful changes, so strange and inexplicable, must

have long perplexed the early student of the heavens. Hitherto, the stars along the moon's route had engaged special attention ; but at length certain bright and conspicuous constellations, towards the north, arrested the eye : and these were watched to see whether they would disappear. Some were found to dip below the western horizon, soon to re-appear in the east ; while others, revolving with the general heavens, rose high above the horizon, swept steadily round, sunk far down, but never disappeared from the sight. This remarkable discovery soon led to another equally important. In watching the stars in the north through an entire night, they all seemed to describe circles ; having a common centre, these circles grew smaller and smaller as the stars approached nearer to the centre of revolution, until finally one bright star was found, whose position was ever fixed ; alone unchanged, while all else was slowly moving. The discovery of this remarkable star must have been hailed with uncommon delight by the primitive observer of the heavens. If his deep devotion to the study of the skies had created surprise among his rude countrymen, when he came to point them to this never changing light hung up in the heavens, and explained its uses in guiding their wanderings on the earth, their surprise must have given place to admiration. Here was the first valuable gift of primitive astronomical science to man.

But, to the astronomer, this discovery opened up a new field of investigation, and light began to dawn on some of the most mysterious questions which had long perplexed him. He had watched the constellations near the moon's track slowly disappear in the effulgence of the sun, and when they were next seen, it was in the east, in the early dawn, apparently emerging from the solar beams, having actually passed by the sun. Watching and reflecting, steadily pursuing the march of the northern constellations, which never entirely disappeared, and noting the relative positions of these, and those falling into the sun, it was at last discovered that the entire starry heavens were slowly moving forward to meet and pass by the sun, or else the sun itself was actually moving backward among the stars. This apparent motion had already been detected in the moon, and now came the reward of long and diligent perseverance. The grand discovery was made that both the sun and moon were moving among the fixed stars, not *apparently*, but *absolutely*. The previously received explanation of the moon's motion could no longer be sustained; for the starry heavens could not at the same time so move as to pass by the moon in one month, and to pass by the sun in a period twelve times as great. A train of the most important conclusions flowed at once from this great discovery. The starry heavens passed beneath and around the earth; the sun and moon were wandering in the same direction, but with different velocities, among the stars; the constellations actually filled the entire heavens above the earth and beneath the earth; the stars were invisible in the day-time, not because they did not exist, but because their feeble light was lost in the superior brilliancy of the sun. The heavens were spherical, and encompassed, like a shell, the entire earth; and hence it was conceived that the earth itself was also a globe, occupying the centre of the starry sphere.

It is impossible for us, familiar as we are at this day with these important truths, to appreciate the rare merit of him who, by the power of his genius, first rose to their knowledge and revealed them to an astonished world. We delight to honour the names of Kepler, of Galileo, of Newton; but here are discoveries so far back in the dim past, that all trace of their origin is lost, which vie in interest and importance with the proudest achievements of any age.

With a knowledge of the sphericity of the heavens, the revolution of the sun and moon, the constellations of the celestial sphere, the axis of its diurnal revolution, astronomy began to be a science, and its future progress was destined to be rapid and brilliant. A line drawn from the earth's centre to the north star formed the axis of the heavens, and day and night around this axis all the celestial host were noiselessly pursuing their never ending journeys. Thus far, the only moving bodies known were the sun and moon. These large and brilliant bodies, by their magnitude and splendour, stood out conspicuously from among the multitude of stars, leaving these minute but beautiful points of light, in one great class, unchangeable among themselves, fixed in their groupings and configurations, furnishing admirable points of reference, in watching and tracing out the wanderings of the sun and moon.

To follow the moon as she pursued her journey among the stars was not difficult; but to trace the sun in his slower and more majestic motion, and to mark accurately his track, from star to star, as he heaved upward to meet the coming constellations, was not so readily accomplished. Night after night, as he sunk below the horizon, the attentive watcher marked the bright stars near the point of setting which first appeared in the evening twilight. These gradually sunk towards the sun on successive nights ; and thus was he traced from constellation to constellation, until the entire circuit of the heavens was performed, and he was once more attended by the same bright stars that had watched, long before, his sinking in the west. Here was revealed the measure of the *Year*. The earth had been verdant with the beauties of spring, glowing with the maturity of summer, rich in the fruits of autumn, and locked in the icy chains of winter, while the sun had circled round the heavens. His entrance into certain constellations marked the coming seasons, and man was beginning to couple his cycle of pursuits on earth with the revolutions of the celestial orbs.

While intently engaged in watching the sun as it slowly heaved up to meet the constellations, some ardent devotee to this infant science at length marked in the early twilight a certain brilliant star closely attendant upon the sun. The relative position of these two objects was noted, for a few consecutive nights, when, with a degree of astonishment of which we can form no conception, he discovered that this brilliant star was rapidly approaching the sun, and actually changing its place among the neighbouring stars ; night after night he gazes on this unprecedented phenomenon, *a moving star !* and on each successive night he finds the wanderer coming nearer and nearer to the sun. At last it disappears from sight, plunged in the beams of the upheaving sun. What had become of this strange wanderer ? Was it lost for ever ? were questions which were easier asked than answered. But patient watching had revealed the fact, that when a group of stars, absorbed into the sun's rays, disappeared in the west, they were next seen in the eastern sky, slowly emerging from his morning beams. Might it not be possible, that this wandering star would pass by the sun, and re-appear in the east ? With how much anxiety must this primitive discoverer have watched in the morning twilight ? Day after day he sought his solitary post, and marked the rising stars, slowly lifting themselves above the eastern horizon. The gray dawn came, and the sun shot forth a flood of light ; the stars faded and disappeared, and the watcher gave over till the coming morning. But his hopes were crowned at last. Just before the sun broke above the

horizon, in the rosy east, refulgent with the coming day, he
descried the pure white silver ray of his long lost wanderer. It
passed the sun; it rose in the east; the first *planet* was
discovered!

With how much anxiety and interest did the delighted
discoverer trace the movements of his wandering star. Here
was a new theme for thought, for observation, for investigation.
Would this first planet sweep round the heavens, as did the sun
and moon? Would it always move in the same direction? Would
its path lie among those groups of stars among which the sun
and moon held their course? Encouraged by past success, he
rejoicingly enters on the investigations of these questions. For
some time the planet pursues his journey from the sun, leaving
it farther and farther behind. But directly it slackens its pace,
it actually stops in its career, and the astonished observer perhaps
thinks that his wandering star had again become fixed. Not so;
a few days of watching dispels this idea. Slowly at first, and
soon more swiftly, the planet seeks again the sun, moving backwards on its former path, until finally its light is but just visible
in the east at early dawn. Again it is lost in the sun's beams for
a time, and, contrary to all preceding analogy, when next seen,
its silver ray comes out, pure and bright, just above the setting
sun. It now recedes from the sun, on each successive evening
increasing its distance, till it again reaches a point never to be
passed; here it stops, is stationary for a day or two, and then
again sinks downward to meet the sun. How wonderful and
inexplicable the movements of this wandering star must have
appeared in the early ages, oscillating backward and forward,
never passing its prescribed limits, and ever closely attendant
upon the sun! Where the sun sank to repose, there did the
faithful planet sink, and where the sun rose, at the same point did
the wandering star makes its appearance. The number of days
was accurately noted, from the stationary point in the east above
the sun, to the stationary point in the west above the sun, and
thus the period, 584 days, from station to station, became known.

The discovery of one planet led the way to the rapid discovery
of several others. If we may judge of their order by their
brilliancy, Jupiter was the second wanderer revealed among the
stars. Then followed Mars and Saturn, and after a long interval
Mercury was detected, hovering near the sun, and imitating the
curious motions of Venus.

Here the progress of planetary discovery was suddenly arrested: keen as was the vision of the old astronomer, long and patient as was his scrutiny, no depth of penetration of unaided vision could stretch beyond the mighty orbit of Saturn, and the search was given over. A close examination of the planets revealed many important facts. Three of them—Mars, Jupiter, and Saturn—were found to perform the circuit of the heavens, like the sun and moon, and in the same direction; with this remarkable difference, that while the sun and moon moved steadily and uniformly in the same direction, the planets occasionally slackened their pace, would then stop, move backwards on their track, stop again, and finally resume their onward motion. Their periods of revolution were discovered by marking the time which elapsed after setting out from some brilliant and well-known fixed star, until they should perform the entire circuit of the heavens, and once more return to the same star. The times of revolution were found to differ widely from each other; Mars requiring about 687 days, Jupiter 4,332 days, and Saturn 10,759 days, or nearly thirty of our years.

The planets all pursued their journeys in the heavens, among the same constellations which marked the paths of the sun and moon, and hence these groups of stars concentrated the greatest amount of attention among the early astronomers, and became distinguished from all the others.

Whatever light may be shed upon antiquity by deciphering the hieroglyphic memorials of the past, there is no hope of ever going far enough back to reach even the nation to which we are indebted for the first rudiments of the science of the stars.

Thus far in the prosecution of the study of the heavens, the eye and the intellect had accomplished the entire work. Rapidly as we have sketched the progress of early discovery, and short as may have been the period in which it was accomplished, no one can fail to perceive how vast is the difference between the light that thus early broke in upon the mind, heralding the coming of a brighter day, and the deep and universal darkness which had covered the world, before the dawn of science. Encouraged by the success which had thus far rewarded patient toil, the mind of man pushes on its investigations deeper and deeper into the domain of the mysterious and unknown.

In watching the annual revolution of the sun among the fixed stars, one remarkable peculiarity had long been recognised.

While the interval of time, from the rising to the setting of the stars, was ever the same at all seasons of the year, the interval from the rising to the setting of the sun was perpetually changing, passing through a cycle which required exactly one year for its completion. It became manifest that the sun did not prosecute its annual journey among the stars, in a circle parallel with those described by the stars, in their diurnal revolution. His path was oblique to those circles ; and while he participated in their diurnal motion, he was sweeping by his annual revolution round the heavens, and was at the same time, by another most extraordinary movement, carried towards the north to a certain distance, then stopping, commenced a return towards the south, reached his southern limit, again changed his direction, and thus oscillated from one side to the other of his mean position.

These wonderful changes became the objects of earnest investigation. In what curve did the sun travel among the stars ? All diurnal motion was performed in a circle, the first discovered, the simplest and the most beautiful of curves ; and in this curve, analogy taught the early astronomers that all celestial movements must be performed. It became therefore a matter of deep interest to trace the sun's path accurately among the stars, to mark his track and to see whether it would not prove to be a circle. To accomplish this, more accurate means must be adopted than the mere watching of the stars which attended the rising or setting sun. The increase and decrease of the shadow of some high pointed rock, to whose refreshing shade the shepherd astronomer had repaired in the heat of noon, and beneath which he had long pondered this important problem, first suggested the means of its resolution. As the summer came on, he remarked that the length of the noon shadow of his rock perpetually decreased from day to day. As the sun became more nearly vertical at noon, the shadow gave him less and less shelter. Watching these noon shadows from day to day, he found them proportioned to the sun's northern or southern motion ; and finally, the thought entered his mind, that these shadows would mark with certainty the limits of the sun's motion north and south, the character of his orbit or route among the stars, the changes and duration of the seasons, and the actual length of the year, which thus far had been but roughly determined. To accomplish the observations more accurately, an area on the ground was smoothed and levelled, and in its centre a vertical pole was erected some ten or fifteen feet in length, whose sharp vertex cast a well defined shadow. And here we have the first astronomical instrument (the gnomon) ever devised by the ingenuity of man. Simple as it is, by its aid the most valuable results were obtained.

The great point was to mark with accuracy the length of the noon-day shadow, from month to month, throughout the entire year. Four remarkable points in the sun's annual track, were very soon detected and marked. One of these occurred in the summer, and was that point occupied by the sun on the day of the shortest noon shadow. Here the sun had reached his greatest northern point, and for a few days the noon shadow cast by the gnomon appeared to remain the same, and the sun *stood still.* The noon shadows now increased slowly, for six months, as the sun moved south, till a second point was noted, when the noon shadow had reached its greatest length. Again it became stationary, and again the sun paused and *stood still,* before commencing his return towards the north. These points were called the *summer* and *winter solstices,* and occurred at intervals of half a year. At the summer solstice the longest day occurred, while at the winter solstice the shortest day was always observed. These extreme differences between the length of the day and night, occasioned the determination of the other two points. From the winter solstice the noon shadows decreased as the length of the day increased, until finally the day and night were remarked to be of *equal* length, and the distance to which the shadow of the gnomon was thrown on that day was accurately fixed. If, on this day, the diurnal circle described by the sun could have been marked in the heavens by a circle of light, sweeping from the east to the west, so that the eye might rest upon and retain it, and if at the same time the sun's annual path among the fixed stars could have been equally exhibited in the heavens by a circle of light, these two circles would have been seen to cross each other, and at their point of crossing, the sun would have been found. The diurnal circle was called the *equator,* the sun's path the *ecliptic,* and the point of intersection was called, appropriately, the *equinox.* As the sun crossed the equator in the spring and autumn, these points received the names of the *vernal* and *autumnal equinoxes,* and were marked with all the precision which the rude means then in use rendered practicable.

The bright circle already imagined in the heavens to represent the sun's annual track among the stars, passed obliquely across the equator, and the amount by which these circles were inclined to each other was actually measured, in these early ages, with no mean precision, by the noon shadows of the gnomon. The ray casting the shortest noon shadow was inclined to the ray forming the longest noon shadow, under an angle precisely double of the inclination of the ecliptic or sun's path to the equator, and the inclination of these two rays marked exactly the annual motion of the sun from south to north, or from north to south. A close examination of the order of increase and decrease in the length of the noon shadows cast by the gnomon, demonstrated the important truth already suspected, that the sun's path was actually a *circle*, but inclined, as has already been shown, to the diurnal circles of the stars and to the equator.

By counting the days which elapsed from the summer solstice to the summer solstice again, a knowledge of the length of the year, or period of the sun's revolution, was obtained. But here again a discovery was made, which produced an embarrassment to the early astronomers, which all their perseverance and research never succeeded in removing.

In these primitive ages, the heavenly bodies were regarded with feelings little less than the reverence we now bestow on the Supreme Creator. The sun especially, as the lord of life and light, was regarded with feelings nearly approaching to adoration, even by the astronomers themselves. The idea early became fixed, that the chief of the celestial bodies must move with a uniform velocity in a circular orbit, never increasing or decreasing: change being inconsistent with the supreme and dignified station which was assigned to him. What then must have been the astonishment of the primitive astronomers, who, in counting the days from the summer to the winter solstice, and from the winter round to the summer solstice, found these intervals to be unequal? This almost incredible result was confirmed, by remarking that the shortest spaces from equinox to solstice, dividing the sun's annual route into four equal portions, were passed over in unequal times. These results could not be doubted, for each observation, from year to year, confirmed them. They were received and recorded, but the problem was handed down to succeeding generations for solution.

In consequence of the oblique direction of the ecliptic, or sun's track, it was found difficult to retain its position in the mind. To

assist in the recurrence to this important circle, a brazen circle
was at length devised, and fastened permanently to another
brazen circle of equal size, under an angle exactly equal to the
inclination of the equator to the ecliptic. Circles, perpendicular
to the equator, and passing through the solstices and equinoxes,
completed the second astronomical instrument, *the sphere*. Having
constructed this simple piece of machinery, it was mounted on an
axis passing through its centre, and perpendicular to its equator,
so as to revolve as did the heavens, whose motions it was intended
to represent. Having so placed the axis of rotation that its prolongation would pass through the north star, this rude sphere
came to play a most important part in the future investigations
of the heavens. Its brazen equator and ecliptic were each divided
into a certain number of equal parts, by reference to which the
motion of the heavenly bodies might be followed with far greater
precision than had ever been previously obtained.

Armed with a new and more perfect instrument, the astronomer
resumes his great investigation. Finding it now possible to mark
out the sun's path in the heavens with certainty by means of his
brazen ecliptic, he discovers that the moon and planets in each
revolution pass across the sun's track, and spend nearly an equal
amount of time on the north and south sides of the ecliptic. This
discovery led to a more accurate determination of the periods of
revolution of the planets. The interval was noted from one
passage across the ecliptic to the next on the same side, and these
intervals marked with accuracy the planetary periods. It now
became possible to fix with greater certainty, the relative positions
of the sun and moon, and problems were once more resumed
which had thus far baffled every effort of human genius. The
phases of the moon, the very first point of investigation, had never
yet yielded up its hidden cause ; and those terrific phenomena,
solar and lunar eclipses, which had long covered the earth with
terror and dismay, were wrapped in mystery, and their explanation had resisted the sagacity of the most powerful and gifted
intellects.

No one has ever witnessed the going out of the sun in dim
eclipse, even now when its most minute phenomena are predicted with rigorous exactitude, without a feeling of involuntary
dismay. What then must have been the effect upon the human
mind in those ages of the world, when the cause was unknown,
and when these terrific exhibitions burst on earth's inhabitants

unheralded and unannounced ? Here then was an investigation, not prompted by curiosity alone, but involving the peace and security of man in all coming ages. We cannot doubt that the causes of the solar eclipse were first detected. It was observed that no eclipse of the sun ever occurred when the moon was visible. Even during a solar eclipse, when the sun's light had entirely faded away, and the stars and planets stole gently upon the sight in the sombre and unnatural twilight, the moon was sought for in vain ; she was never to be seen. This fact excited curiosity, and gave rise to a careful and critical examination of the place in which the moon should be found immediately after a solar eclipse ; and it was soon discovered that on the night following the day of eclipse, the moon was seen in her crescent shape very near to the sun and but a short distance from the sun's path. By remarking the moon's place, next before a solar eclipse, and that immediately following, it was seen that at the time of the occurrence of the eclipse, the moon was actually passing from the west to the east side of the sun's place, and finally a little calculation showed that a coincidence of the sun and moon in the heavens took place at the precise time at which the sun had been eclipsed. The conclusion was irresistible, and the great fact was announced to the world, that the sun's light *was hidden by the interposition of the dark body of the moon.*

Having reached this important result with entire certainty, the explanation of the moon's phases followed in rapid succession. For it now became manifest that the moon shone with *borrowed light,* and that her brilliancy came from the reflected beams of the sun. This was readily demonstrated by the following facts. When the moon was so situated that the side next to the sun (the illuminated one) was turned from the eye of the observer, (as was the case in a solar eclipse,) then the moon's surface next to the observer was always found to be entirely black. Pursuing her journey from this critical point, the moon was next seen near the sun in the evening twilight, as a slender thread of light, a very small portion of her illuminated surface being now visible. Day after day this visible portion increases, until finally the moon rises as the sun sets, full orbed and round, being directly opposite the sun, and turning her entire illuminated surface towards the eye of the observer. By like degrees she loses her light as she approaches, and finally becomes invisible as she passes by the sun. From this examination it became evident

D

that the moon was a globular body, non-luminous, and revolving in an orbit, comprehended entirely within that described by the sun, and consequently nearer to the earth than the sun. Having ascertained this fact, it was concluded that among all the moving heavenly bodies, the periods of revolution indicated their relative distances from the earth. Hence Mars was regarded as more distant than the sun, Jupiter more remote than Mars, and Saturn the most distant, as it was the slowest moving of all the planets.

After reaching to a knowledge of the causes producing the eclipses of the sun and the phases of the moon, it remained yet to resolve the mystery of the lunar eclipse. It was far more difficult to render a satisfactory account of this phenomenon than either of the preceding. The light of the moon was not intercepted by the interposition of any opaque body, between it and the eye of the observer. No such body existed, and long and perplexing was the effort to explain this wonderful phenomenon. Finally, it was observed that all opaque bodies cast shadows in directions opposite to the source of light. Was it not possible that the light of the sun, falling upon the earth, might be intercepted by the earth, and thus produce a shadow which might even reach as far as the moon! So soon as this conjecture was made, a series of examinations were commenced to confirm or destroy the theory. It was at once seen, that in case the conjecture was true, no lunar eclipse could occur except when the sun, earth, and moon were situated in the same straight line ; a position which could never occur except at the *full* or *new* of the moon. It was soon discovered that it was only at the full that lunar eclipses took place, thus confirming the truth of the theory, and fixing it beyond a doubt, that the shadow of the earth falling on the moon was the cause of her eclipse. The moon had already been shown to be non-luminous ; and the moment the interposition of the earth between it and its source of light, the sun, cut off its light, it ceased to be visible, and passed through an eclipse. The sphericity of the earth, which had been analogically inferred from that of the heavens, was now made absolutely certain; for it was remarked, as the moon entered the earth's shadow, that the track of this dark shadow across the bright surface of the moon was always *circular*, which was quite impossible for every position, except the earth which cast this circular shadow should be of a globular form.

Having now attained to a clear and satisfactory explanation of the two grand phenomena, solar and lunar eclipses, the question naturally arose, Why was not the sun eclipsed in each revolution of the moon ? and how happened it that the moon in the full did not always pass through the earth's shadow? An examination of the moon's path among the fixed stars gave to these questions a clear and positive answer. It was found that the sun and moon did not perform their revolutions in the same plane. The moon's route among the stars crossed the sun's route under a certain angle, and it thus frequently happened, that at the new and full, the moon occupied some portion of her orbit too remote from that of the sun to render either a lunar or solar eclipse possible.

Rapidly have we traced the career of discovery. The toil and watching of centuries have been condensed into a few moments of time, and questions requiring ages for their solution have been asked only to be answered. In connection with the investigations just developed, and as a consequence of their successful prosecution, the query arose, Whether in case science had reached to a true exposition of the causes producing an eclipse of the sun, was it not possible to stretch forward in time, and anticipate and predict the coming of these dread phenomena ?

To those who have given but little attention to the subject, even in our own day, with all the aids of modern science, the prediction of an eclipse seems sufficiently mysterious and unintelligible. How then it was possible, thousands of years ago, to accomplish the same great object without any just views of the structure of the system, seems utterly incredible. Follow me then, while I attempt to reveal the train of reasoning which led to the prediction of the first eclipse of the sun, the most daring prophecy ever made by human genius. Follow in imagination this bold interrogator of the skies to his solitary mountain summit, withdrawn from the world, surrounded by his mysterious circles, there to watch and ponder through the long nights of many, many years. But hope cheers him on, and smooths his rugged pathway. Dark and deep is the problem ; he sternly grapples with it, and resolves never to give over till victory crown his efforts.

He has already remarked that the moon's track in the heavens crossed the sun's, and that this point of crossing was in some way intimately connected with the coming of the dread eclipse.

D 2

He determines to watch and learn whether the point of crossing
was fixed, or whether the moon in each successive revolution
crossed the sun's path at a different point. If the sun in its
annual revolution could leave behind him a track of fire marking
his journey among the stars, it is found that this same track was
followed from year to year, and from century to century, with
undeviating precision. But it was soon discovered that it was
far different with the moon. In case she too could leave behind
her a silver thread of light sweeping round the heavens, in completing one revolution, this thread would not join, but would
wind around among the stars in each revolution, crossing the
sun's fiery track at a point west of the previous crossing. These
points of crossing were called the *moon's nodes.* At each revolution the node occurred farther west, until after a circle of about
nineteen years, it had circulated in the same direction entirely
round the ecliptic. Long and patiently did the astronomer-watch
and wait ; each eclipse is duly observed, and its attendant circumstances are recorded ; when at last the darkness begins to
give way, and a ray of light breaks in upon his mind. He finds
that no eclipse of the sun ever occurs unless the *new moon is in
the act of crossing the sun's track.* Here was a grand discovery.
He holds the key which he believes will unlock the dread mystery, and now, with redoubled energy, he resolves to thrust it
into the wards and drive back the bolts.

To predict an eclipse of the sun, he must sweep forward from
new moon to new moon, until he finds some new *moon* which should
occur while the moon was in the act of crossing from one side to
the other of the sun's track. This certainly was possible. He knew
the exact period from new moon to new moon, and from one
crossing of the ecliptic to another. With eager eye he seizes the
moon's place in the heavens, and her age, and rapidly computes
where she will be at her next change. He finds the new moon
occurring far from the sun's track ; he runs round another
revolution ; the place of the new moon falls closer to the sun's
path, and the next yet closer, until, reaching forward with
piercing intellectual vigour, he at last finds a new moon which
occurs precisely at the computed time of her passage across the
sun's track. Here he makes his stand, and on the day of the
occurrence of that new moon, he announces to the startled inhabitants of the world that the sun shall expire in dark eclipse !
Bold prediction ! Mysterious prophet ! with what scorn must

the unthinking world have received this solemn declaration!
How slowly do the moons roll away, and with what intense
anxiety does the stern philosopher await the coming of that day
which should crown him with victory, or dash him to the ground
in ruin and disgrace. Time to him moves on leaden wings; day
after day, and at last hour after hour, roll heavily away. The
last night is gone; the moon has disappeared from his eagle gaze
in her approach to the sun, and the dawn of the eventful day
breaks in beauty on a slumbering world.

This daring man, stern in his faith, climbs alone to his rocky
home, and greets the sun as he rises and mounts the heavens,
scattering brightness and glory in his path. Beneath him is
spread out the populous city, already teeming with life and
activity. The busy morning hum rises on the still air, and
reaches the watching place of the solitary astronomer. The
thousands below him, unconscious of his intense anxiety, buoyant
with life, joyously pursue their rounds of business, their cycles
of amusement. The sun slowly climbs the heaven, round and
bright and full orbed. The lone tenant of the mountain top
almost begins to waver in the sternness of his faith as the morning hours roll away. But the time of his triumph, long delayed,
at length begins to dawn; a pale and sickly hue creeps over the
face of nature. The sun has reached his highest point, but his
splendour is dimmed, his light is feeble. At last it comes!
Blackness is eating away his round disc; onward with slow but
steady pace the dark veil moves, blacker than a thousand nights;
the gloom deepens; the ghastly hue of death covers the universe;
the last ray is gone, and horror reigns! A wail of terror fills
the murky air, the clangour of brazen trumpets resounds, an
agony of despair dashes the stricken millions to the ground;
while that lone man, erect on his rocky summit, with arms out-
stretched to heaven, pours forth the grateful gushings of his
heart to God, who had crowned his efforts with triumphant victory. Search the records of our race, and point me, if you can,
to a scene more grand, more beautiful. It is to me the proudest
victory that genius ever won. It was the conquering of nature,
of ignorance, of superstition, of terror, all at a single blow, and
that blow struck by a single arm. And now do you demand the
name of this wonderful man? Alas! what a lesson of the
instability of earthly fame are we taught in this simple recital.
He who had raised himself immeasureably above his race, who

must have been regarded by his fellows as little less than a god, who had inscribed his fame on the very heavens, and had written it in the sun, with a "pen of iron, and the point of a diamond," even this one has perished from the earth; name, age, country, are all swept into oblivion. But his proud achievement stands. The monument reared to his honour stands; and although the touch of time has effaced the lettering of his name, it is powerless, and cannot destroy the fruits of his victory.

A thousand years roll by: the astronomer stands on the watch-tower of old Babylon, and writes for posterity the records of an eclipse; this record escapes destruction, and is safely wafted down the stream of time. A thousand years roll away; the old astronomer, surrounded by the fierce, but wondering Arab, again writes, and marks the day which witnesses the sun's decay. A thousand years roll heavily away: once more the astronomer writes from amidst the gay throng that crowds the brightest capital of Europe. Record is compared with record, date with date, revolution with revolution, the past and present are linked together; another struggle commences, and another victory is won. Little did the Babylonian dream that he was observing for one, who, after the lapse of 3000 years, should rest upon this very record the successful resolution of one of nature's darkest mysteries.

We have now reached the boundary where the stream of discovery, which we have been tracing through the clouds and mists of antiquity, begins to emerge into the twilight of tradition, soon to flow on in the clear light of a history that shall never die. Henceforth our task will be more pleasing, because more certain; and we invite you to follow us as we attempt to exhibit the coming struggles and future triumphs of the student of the skies.

LECTURE III.

F in tracing the career of astronomy in the primitive ages of the world, we have been left to pursue our way dimly, through cloud and darkness ; if regrets rise up, that time has swept into oblivion the names and country of the early discoverers ; in one reflection there is some compensation : while the bright and enduring truths which they wrested from nature have descended to us, their errors, whatever they may have been, are for ever buried with their names and their persons. We are almost led to believe that those errors were few and transient, and that the mind, as yet undazzled by its triumphs, questioned nature with that humility and quiet perseverance which could bring no response but truth.

In pursuing the consequences flowing from the prediction of an eclipse, several remarkable results were reached, which we proceed to unfold. It will be recollected that to produce either solar or lunar eclipses, the new or full moon must be in the act of crossing the sun's annual track. This point of crossing, called the *moon's node*, became therefore an object of the deepest interest. Long and careful scrutiny revealed the fact of its movement around the ecliptic, in a period of eighteen years and eleven days, during which time there occur 223 new moons, or 223 full moons. If then, a new moon falls on the sun's track to produce a solar eclipse to-day, at the expiration of 223 lunations, again will the new moon fall on the ecliptic, and an eclipse will surely take place. Suppose, then, that all the eclipses which occur within this remarkable period of 223 lunations are carefully observed, and the days on which they fall recorded, on each and every one of these days, during the next period of 223 lunations, eclipses may be expected, and their coming foretold.

This wonderful period of eighteen years and eleven days, or 223 lunations, was known to the Chaldeans, and by its use eclipses were predicted, more than 3000 years ago. It is likewise found among the Hindoos, the Chinese, and the Egyptians, nations widely separated on the earth's surface, and suggesting the idea that it had its origin among a people even anterior to the Chaldeans. It is now known by the name of the *Zaros*, or Chaldean period.

Let it not be supposed that the application of the Zaros to the prediction of eclipses, can in any way supersede modern methods. While antiquity contented itself with announcing the *day* on which the dark body of the moon should hide the sun, modern science points to the exact *second* on the dial, which shall mark the first delicate contact of the moon's edge with the brilliant disc of the sun.

It would be a matter of great interest to fix the epoch of primitive discovery. Though this is impossible, its high antiquity is attested by a few facts, to which we will briefly advert. We find among all the ancient nations, Chaldeans, Persians, Hindoos, Chinese, and Egyptians, that the seven days of the week were in universal use, and what was far more remarkable, each of these nations named the days of the week after the seven planets, numbering the sun and moon among the planets. It is, moreover, found that the order of naming is not that of the distance, velocity, or brilliancy of the planets, and neither does the first day of the week coincide among the different nations; but the order once commenced is invariably preserved by all. If we compute the probability of such a coincidence resulting by accident, we shall find the chances millions to one against it. We are, therefore, forced to the conclusion, that the planets were discovered, and the seven days of the week devised and named, by some primitive nation, from whom the tradition descended imperfectly, to succeeding generations.

A remarkable discovery, made in the remote ages of the world, throws some farther light on the era of the primitive astronomical researches. The release of the earth from the icy fetters of winter, the return of spring, and the revivification of nature, is a period hailed with uncommon delight, in all ages of the world. To be able to anticipate its coming, from some astronomical phenomenon, was an object of earnest investigation by the ancients.

It was found that the sun's entrance into the equinox, reducing to equality the length of the day and night, always heralded the coming of the spring. Hence to mark the equinoctial point among the fixed stars, and to note the place of some brilliant star whose appearance in the early morning dawn would announce the sun's approach to the equator, was early accomplished with all possible accuracy. This star once selected, it was believed that it would remain for ever in its place.

The sun's path among the fixed stars had been watched with success, and it seemed to remain absolutely unchanged; and hence the points in which it crossed the equator, for a long while, were looked upon as fixed and immoveable. And, indeed, centuries must pass away before any change could become sensible to the naked eye and its rude instrumental auxiliaries. But a time arrives at last when the bright star, which for more than five hundred years had with its morning ray announced the season of flowers, is lost. It has failed to give its warning; spring has come, the forests bud, the flowers bloom, but the star which once gave promise, and whose ray had been hailed with so much delight by many generations, is no longer found. The hoary patriarch recals the long experience of a hundred years, and now perceives that each succeeding spring had followed more and more rapidly after the appearance of the sentinel star. Each year the interval from the first appearance of the star in the early dawn, up to the equality of day and night, had grown less and less, and now the equinox came, but the star remained invisible, and did not emerge from the sun's beams until the equinox had passed.

Long and deeply were these facts pondered and weighed. At length truth dawned, and the discovery broke upon the unwilling mind, that the sun's *path among the fixed stars was actually changing,* and that his point of crossing the equator was slowly moving backwards towards the west, and leaving the stars behind. The same motion, only greatly more rapid, had been recognised in the shifting of the moon's node, and in the rapid motion of the points at which her track crossed the equator. The retrograde motion of the equinoctial points caused the sun to reach these points earlier than it would have done had they remained fixed, and hence arose the *precession* of the *equinoxes.*

This discovery justly ranks among the most important achieved by antiquity. Its explanation was infinitely above the reach of human effort at that early day; but to have detected the fact, and to have marked a motion so slow and shrouded, gives evidence of a closeness of observation worthy of the highest admiration. It will be seen hereafter, that the human mind has reached to a full knowledge of the causes producing the retrograde movement of the equinoxes among the stars. Its rate of motion has been determined, and its vast period of nearly twenty-six thousand years has been fixed. Once revealed, the slow movement of the equinox makes it a fitting hour-hand, on the dial of the heavens, with which to measure the revolutions of ages. As the sun's path has been divided into twelve constellations, each filling the twelfth part of the entire circuit of the heavens, for the equinox to pass the twelfth part of the dial, or from one constellation to the next, will require a period of more than two thousand years. Since the astronomer first noted the position of this hour-hand on the dial of the stars, but one of its mighty hours of two thousand years has rolled away. In case any record could be found, any chiselled block of granite, exhibiting the place of the equinox among the stars at its date, no matter if ten thousand years had elapsed, we can reach back with certainty, and fix the epoch of the record.

No such monument has ever been found; but there are occasional notices of astronomical phenomena, found among the Greek and Roman poets, which at least give colour to conjecture. Virgil informs us that the "White Bull opens with his golden horns this year:"—

> "Candidus auratis aperit cum cornibus annum,
>   Taurus."

This statement we know is not true, if applied to the age in which the poet wrote, and seems to be the quotation of an ancient tradition. If this conjecture be true, this tradition must have been carried down the stream of time for more than two thousand years, to reach the age in which the poet wrote. Although these conjectures are vague and uncertain, the frequent allusions to the constellations of the zodiac in the old Hebrew Scriptures, and in the works of all ancient writers, sufficiently attest the extreme antiquity of these arbitrary groupings of the stars.

In taking leave of the primitive ages of astronomy, and in entering on that portion of the career of research and discovery whose history has been preserved, let us pause for a moment and consider the position occupied by the human mind at this remarkable epoch.

Thus far the eye had done its work faithfully. Through long and patient watching, it had revealed the facts, from which reason had wrought out her great results. The stars grouped into constellations, glittered in the blue concave of a mighty sphere, whose centre was occupied by the earth. Within this hollow sphere, sun, moon, and planets, kept their appointed courses, and performed their ceaseless journeys. Their wanderings had been traced; their pathway in the heavens was known, their periods determined, the inclinations of their orbits fixed. So accurately had the eye followed the sun and moon, that it had learned to anticipate their relative positions, their oppositions and conjunctions, till, reaching forward, it had robbed the dread eclipse of its terrors, and had learned to hail its coming with delight. The pathway of the sun and moon among the stars had been scanned and studied, until their slowest changes had been marked and measured.

Such were the rich fruits of diligence and perseverance which descended from the remote nations of antiquity. With the advantage of these great discoveries, and the experience of preceding ages, it is natural to expect rapid progress, when science found its home among the bold, subtle, and inquisitive Greeks. He who entertains this expectation will meet with disappointment. Not that investigations were less constantly or perseveringly conducted; not that less perfect means were employed, or less powerful talent consecrated to the work; but because a point had been reached of exceeding difficulty. The era of discovery from mere inspection was rapidly drawing to a close. It was an easy matter to count the days from full moon to full moon, to watch a planet as it circled the heavens from a fixed star until it returned to the same star again, to mark its stopping, its reversed motion, and its onward goings; but it was a far different matter to rise to a knowledge of the causes of these stations and retrogradations, and to render a clear and satisfactory account of them. The problem now presented, was to combine all the facts treasured by antiquity, all the movements exhibited in the heavens, and reduce them to simplicity and harmony. The Greek philosophers, from Plato down to the extinction of the last school of philosophy, recognised this to be the true problem, and essayed its solution, with an energy and pertinacity worthy of the highest admiration.

Let us now examine the causes which arrested the progress of astronomical discovery, and held back the mind for a period of more than two thousand years. Surrounded as we are by the full blaze of truth, accustomed to the simplicity and beauty which now reign everywhere in the heavens, we find it next to impossible to realise the true position of those brave minds, which, enveloped in darkness, deceived by the senses, fettered by prejudice, struggled on and finally won the victory, whose fruits we enjoy.

The most careful and philosophical examination of the heavens seemed to lead to the admitted truth, that the earth was the centre of all celestial motion. In the configuration of the bright stars there was no change. From age to age, from century to century, immoveably fixed in their relative positions, they had performed their diurnal revolutions around the earth. They were even of the same magnitude, of the same brilliancy. How impossible was this, on any hypothesis, except that of the fixed central position of the earth. Leaving the fixed stars, an examination of the motions of the sun and moon, their nearly uniform velocity, their invariable diameters in all portions of their orbits, demonstrated the central position of the earth with reference to them. To shake a faith thus firmly fixed, sustained by the evidence of the senses, consonant with every feeling of the mind, accordant with fact and reason, required a depth of research, and the development of new truths, only to be revealed after centuries of observation.

Every effort, then, to explain the celestial phenomena, started with the undoubted fact, that the earth was the centre of all motion. Thus far, the mind had not reached the idea of apparent motion. If the moon moved, so equally did the sun. There was exactly the same amount of evidence to demonstrate the reality of the one motion, as the other; neither were doubted. It would have been unphilosophical to reject the one without rejecting the other.

The centre of motion once determined, the nature of the curve described was so obviously presented to the eye, that it seemed impossible to hesitate for one moment. The circle was the only regular curve known to the ancients. Its simplicity, its beauty, and perfection, would have induced its selection, even had there been a multitude of curves from which to choose. Its curvature was ever the same. It had neither beginning nor end. It was the symbol of eternity, and admirably shadowed forth the eternity of the motions to which it gave form. As if these considerations had required confirmation, every star and planet, the sun and moon, all described circles in their diurnal revolution; and it seemed impossible to doubt that their orbitual motions were performed in the same beautiful curve. In truth, observation confirmed this conjecture; and the orbits of all the moving bodies, when projected on the concave heavens, were circles. That this curve, then, should have been adopted without doubt or hesitation, is not to be wondered at. It came therefore to be a fixed principle, that in all hypotheses devised to explain the phenomena of the heavens, circular motion and circular orbits alone could be employed.

To these great principles of the central position of the earth, and the circular orbits, we must add that of the earth's immobility. This doctrine was undoubtedly sustained by the evidence of all the senses which could give testimony. No one had seen it move, had heard it move, had felt it move. How was it possible to doubt the evidence of the eye, the touch, the ear? Here, then, was another incontrovertible fact, which even the most sceptical could not doubt, and which lay at the foundation of all effort to resolve the problem under examination.

With a full knowledge and appreciation of these facts, we are prepared to enter upon an examination of the career of astronomy up to the time when all darkness disappeared before the dawning of a day which should never end. The early Greek philosophers, little fitted by nature for close and laborious observation, rather chose to gather in travel the wisdom which was garnered up in the temples, and among the priests of Egypt and India. Returning to their native country, they theorised on the facts they had learned, and taught doctrines, which found their only support in trains of fanciful or specious reasoning. Thus we find Pythagoras mingling the great discoveries of antiquity with theories the most vague and visionary. While gleams of truth flash occasionally through the darkness of his doctrines, they seem but fortunate guesses. His views were sustained by no solid argument, and rapidly sunk into forgetfulness. This philosopher is said to have fixed the sun in the centre of his planetary system, and to have taught the revolution

Plato.

of the earth in an orbit ; but to sustain this bold conjecture, the only reason assigned, was, that fire, which composes the sun, was more dignified than earth, and hence should hold the more dignified position in the centre. We are not surprised that Hipparchus and Ptolemy, the true astronomers among the Greeks, should have rejected a doctrine sustained by so futile and absurd a reason. Nicetas, a follower of Pythagoras, is said to have gone farther than his master, and to have adopted the idea that the revolution of the heavens was an appearance produced by an actual rotation of the earth on an axis, once in twenty-four hours. This extraordinary and almost prophetic announcement, unfortunately was not sustained by any solid argument. It was regarded as a vain dream, and soon was lost in oblivion.

A crowd of theoretic philosophers filled for a long time the schools of Greece, contributing little to science, and diverting the mind from the only train of research which could lead to any true results. At length a philosopher arose who restored investigation to its legitimate channel. Hipparchus, abandoning for the present all vain effort to explain the phenomena of the heavens, gave himself up to close, continuous, and accurate observation. He began with the movements of the sun in his annual orbit. By the construction of superior brazen circles, he measured the daily motion of the sun during the entire year. He confirmed the discovery of the ancients, of the irregular or unequal progress of this luminary, and fixed that point in the sun's orbit where it moved with greatest velocity. Year after year did this devoted astronomer follow the sun, until finally he discovered that the point on the orbit, where its motion was swiftest, did not remain fixed, but was advancing in each revolution, at a very slow rate along the orbit. Having thus demonstrated and characterised the irregularity of the sun's motion, he directed his attention to minute examinations of the moon, and reached results precisely similar. From these discoveries it became manifest, that in case the motions of the sun and moon were circular and uniform, the earth did not occupy the exact centres of their orbits ; for on this hypothesis any irregularity of motion would have been impossible. Here was a point gained. The exact central position of the earth was disproved in two instances, and even the amount of its eccentricity, or distance from the true centre, determined. Retaining the circular and uniform motion of the sun and moon, the discovered irregularities were tolerably well represented by the eccentric position of the earth, from whose surface these motions were measured.

While pursuing these important researches, Hipparchus resolved upon a work of extraordinary difficulty, which had never before been attempted, and which fully attests the grandeur and sagacity of his views. This enterprise was nothing less than numbering the stars and fixing their positions in the heavens. This he actually accomplished ; and his catalogue of 1081 of the principal stars is, perhaps, the richest treasure which the Greek school has transmitted to posterity. We cannot too much admire the disinterested devotion to science which prompted this great undertaking, and the firmness of purpose which sustained the solitary observer, through long years of toil. It was a work for posterity, and could yield to its author no reward during his life. Conscious of this, his resolution never faltered, and grateful posterity crowns his memory with the well-earned title of Father of Astronomy. The noble example thus set by Hipparchus was not lost on Ptolemy, justly the most distinguished among his immediate successors. An ardent student, a close observer, a patient and candid reasoner, Ptolemy collected and digested the discoveries and theories of his predecessors, and transmitted them, in connection with his own, successfully to posterity. Rejecting the absurd doctrine of the solid crystal spheres of Eudoxus, and the unsustained notions of Pythagoras, this bold Greek undertook the resolution of the great problem, which Plato had long before presented, and to accomplish which so many unsuccessful efforts had been made.

After a careful examination of all the facts and discoveries which the world then possessed, adding his own extensive observations, Ptolemy promulged a system which bears his name, and which endured for more than fourteen hundred years. He fixed the earth as the great centre, about which the sun, the moon, the planets, and the starry heavens, revolved. Retaining the doctrine of uniform circular motion, he accounted for the irregularity in the movements of the sun and moon by the eccentric position of the earth in their orbits. To explain the anomalous movement of the planets, he devised the system of cycles and epicycles. Every planet moved uniformly in the circumference of a small circle, whose centre moved uniformly in the circumference of a large circle, near whose centre the earth was located. By this ingenious theory, it was shown that a planet moving in the circumference of its small circle might appear to retrograde, to become stationary, and finally to advance among the fixed stars. Thus were all the phenomena known to the Greek astronomer so satisfactorily accounted for, that it even became possible from this singular theory, to compute tables of the planetary motions, from which their places could be predicted with such precision, that the error, if any existed, escaped detection by the rude instruments then in use.

While the explanation of the celestial phenomena had constituted the principal object of the Greek astronomers, some rude efforts were commenced to determine the magnitude of the earth, and the relative distances of the sun and moon. The process adopted by Eratosthenes, two thousand years ago, to determine the circumference of the earth and its diameter, is essentially the same now employed by modern science. The results reached by the Greek astronomer, owing to an ignorance of the exact value of his unit, are lost to the world.

When astronomy was banished from Greece, it found a home among the Arabs. When darkness and gloom wrapped the earth through ten long centuries, and human knowledge languished, and art died, and genius slumbered, it is a remarkable fact that astronomy, during that long period of ignorance, instead of being lost, was actually slowly advancing ; and when the dawn of learning once more broke on Europe, the astronomy of the Greeks, improved by the Arabs and the Persians, was preserved in the great work of Ptolemy, and transmitted to posterity.

It is true that no change had been wrought in the Greek theory, but observations had been multiplied and slow changes measured, which prepared the way for the discoveries which were soon to succeed. On the revival of learning in Europe, the literature and science of the Greeks and Romans rapidly spread, and gained an astonishing ascendancy over the human mind. Indeed, theirs was the only science, the only wisdom. Time-honoured, and venerable with age, the philosophy of Aristotle, the geometry of Euclid, and the astronomy of Ptolemy, filled the colleges and universities, and fastened itself upon the age with a tenacity which permitted no one to question or doubt, and which seemed to defy all further progress. Such was the state of science and the world, when Copernicus consecrated his genius to the examination of the heavens.

Copernicus.

To a mind singularly bold and penetrating, Copernicus united habits of profound study and severe observation. Deeply read in the received doctrines of science, he examined with the keenest interest every hint which the philosophers of antiquity had left on record concerning the system of nature. For more than thirty years he watched, with unceasing perseverance, the movements of the heavenly bodies. By the construction of superior instruments, he compared the observed places of the sun, moon, and planets, with their positions computed from the best tables founded on the theory of Ptolemy. The hypothesis of uniform circular motion had originally been adopted to preserve

E

the simplicity of nature, and with true philosophy. But as one
irregularity after another had been discovered in the movements
of the heavenly bodies, each of which must be explained on the
circular hypothesis, one circle had been successively added to
another, eccentrics and epicycles, equants and differents, until, to
preserve simplicity, the system had grown to the most extravagant complexity. The primitive idea of simplicity was a just one,
founded in nature and adopted in reason. But after thirty years
of vain effort to harmonise the phenomena of the heavens with the
theory of Ptolemy, after entangling himself in a maze of complexity in his effort to preserve simplicity, Copernicus was at last
driven to doubt, and doubt soon grew into disbelief. By a close
examination of the motions of Mercury and Venus, he found
that these planets always accompanied the sun, participated in
its movements, and never receded from it except to limited distances. The uniformity of their oscillations, from the one side
to the other of the sun, suggested their revolution about that
luminary, in orbits whose planes passed nearly through the eye
of the observer. The Egyptians had reached to this doctrine,
had communicated it to Pythagoras, who taught it to his
countrymen, nearly two thousand years before the time of
Copernicus.

If, then, simplicity imperiously demanded the abandonment of
the earth as the great centre of motion, in the search for a new
centre, a multitude of circumstances pointed to the sun. It was
the largest and most brilliant of all the heavenly bodies. It gave
light to the moon and planets. It gave life to the earth and its
inhabitants. It was certainly accompanied by two satellites,
and, above all, it was so related to the earth, that if motion in
the one was abandoned, it must instantly, and without a
moment's hesitation, be transferred to the other. Long did the
philosopher hesitate, perplexed with doubts, surrounded by prejudice, embarrassed with difficulties ; but, finally, rising superior
to every consideration save truth, he quitted the earth, swept
boldly through space, and planted himself upon the sun. With
an imagination endowed with the most extraordinary tenacity,
he carried with him all the phenomena of the heavens, which
were so familiar to his eye while viewed from the earth. A long
train of investigation was now before him. He commences with
his now distant earth. Its immobility is gone ; he beholds it
sweeping round the heavens in the precise track once followed

NORTHERN AND SOUTHERN HEMISPHERES OF THE PLANET MARS. (Page 51.)

by the sun. The same constellations mark its career, the same periodic time, the same inequalities of motion; all that the sun has lost the earth has gained.

Thus far, the change had been without results. He now gives his attention to the planets. Here a most beautiful scene broke upon his senses. The complex wanderings of the planets, their stations, their retrograde motions, all disappeared, and he beheld them sweeping harmoniously around him. The earth, deprived of her immobility, started in her orbit, joined her sister planets, and gave perfection to the system. The oscillations of Mercury and Venus were converted into regular revolutions, still holding their places nearest to the sun; then came the earth, next Mars and Jupiter, and last of all Saturn, away in the distance, slowly pursuing his mighty orbit. All were moving in the same direction, their paths filling the same belt of the heavens.

Charmed with this beautiful scene, the philosopher turns to an examination of the moon. Was she, too, destined to take her place among the planets? A short investigation revealed her true character. She could not be a planet revolving about the sun *interior* to the earth's orbit, for, if so, she would have imitated the oscillations of Mercury and Venus. She was not a planet revolving around the sun, *exterior* to the orbit of the earth; for, in that case, she must have imitated the stations and retrogradations of Mars, Jupiter, and Saturn. The invariability of her diameter, as seen from the earth, joined to these considerations, established the fact of her secondary character; and, like a favourite minister who accompanies his dethroned monarch in his exile, so did the faithful moon cling to the earth, and follow it in its wanderings through space.

Such is the beautiful system wrought out by the great Polish philosopher. Far from perfect, it was founded in truth; and, although improvement might and must come, revolution could never shake its firm foundation. While the more prominent irregularities in the planetary motions were removed by constituting the sun the centre of motion, there yet remained an increase and decrease in the orbital velocities of all the planets, now including the earth among the number, which were inexplicable. The planets did not revolve, then, in circles whose exact centre was occupied by the sun. The moon's orbit was not a circle whose exact centre was the earth; and to explain these unfortunate irregularities, Copernicus, clinging to circular motion,

E 2

as the world had done for 2000 years, was driven to adopt the same expedients which characterised the theories of Ptolemy : the eccentric and epicycle were fastened upon the new system of astronomy. Yet another difficulty embarrassed the mind of Copernicus. In giving to the earth a·rotation on its axis once in twenty-four hours, he explained the apparent revolution of the starry heavens. This axis of rotation, it was readily seen, must ever remain parallel to itself in the annual revolution of the earth in its orbit. Being in this way carried round such a vast circumference, the prolongation of the axis ought to pierce the northern heavens in a series of points which would form a curve so large as not to escape detection. But no such curve appeared ; the north pole of the heavens, scrutinised with the most delicate instruments, preserved its position immoveably throughout the entire revolution of the earth in its orbit ; and to escape from this difficulty there was no alternative but to admit that the distance of the sphere of the fixed stars was so great that the diameter of the earth's orbit, equal to 200,000,000 of miles, was absolutely nothing, when compared with that mighty distance.

Under these circumstances, it is not wonderful that Copernicus should have promulged his system with extreme diffidence, and only after long delay ; indeed, his great work, setting forth his doctrines, was never read by its author in print, and only reached him in time to cheer his dying moments.

We cannot, then, be surprised that the new system was received with doubt and distrust, or rather that it was for a long while absolutely rejected. The progress of truth is ever slow, while error moves with rapid pace. The reason is obvious : error is seized by a class of minds, which asks no evidence ; while the searchers for truth adopt it only after the most deliberate examination.

But the revolution had been commenced. A few bold minds were struck with the simplicity and beauty of the conjectures of Copernicus ; and when the exigencies of the age demand genius, it seems to rise spontaneously. The mind had persevered in a system founded in reason, and which nothing short of this very perseverance could have demonstrated to be erroneous. Like the traveller who is uncertain which of two roads to take, he reflects, reasons, and decides, and, even if his choice be a wrong one, it would be folly to stop before fully convinced that he had chosen erroneously.

But the mind is once again in the path of truth ; and after wandering twenty long centuries in darkness, which grew deeper and deeper, the change from darkness to light gives vigour to its movements, and its future achievements are destined to be rapid and glorious.

Here let us pause for a moment on the boundary which divides ancient from modern science, and glance at the collateral circumstances which were found to modify and retard the investigations which had commenced. The old doctrines of philosophy and astronomy had become intimately interwoven with human society. Ptolemy, and Plato, and Aristotle, were regarded with a sort of reverential awe. Even the Church, not following, but leading the world in this profound respect for ancient philosophy, pronounced the doctrines of Ptolemy in accordance with the revelations of Scripture, and girdled them with the fires of persecution, through which alone their sacredness could be attacked. Thus entrenched and defended by prejudice, by society, and by religion, none but the most daring spirit would enter the conflict against such unequal odds. Conscious of these difficulties, Copernicus had wisely avoided collision, and gave his doctrines to the world with such caution as not to provoke attack. But this armed neutrality could not long endure. If the new doctrine were founded in error, left to itself it would never advance, and would soon quietly sink into oblivion. On the contrary, should it prove to be based upon truth, no power could arrest its progress, or stay its development. The contest must come, sooner or later, and demanded in those who should battle for the truth the rarest qualities.

Copernicus had merely commenced the examination of his bold conjecture. A lifetime was too short to accomplish more. He had transferred the centre of motion from the earth to the sun, and rested the truth of his hypothesis on a *diminished* complexity in the celestial phenomena. In case the true centre had been found, it now remained to determine the exact curves in which the planets revolved, the laws regulating their motion, and the nature of the bond which it was now suspected united the planetary worlds into one great system. The resolution of these profound questions was reserved for *Kepler*, who has, without flattery, been termed the Legislator of the heavens, and who has earned the reputation of being *first* in fact and first in genius among modern astronomers. He united, in the most perfect manner, all the qualifications of a great discoverer. Ardent, enthusiastic, and subtle, he pursued his investigations with a keen and restless activity. Pati-

Kepler.

ent, laborious, and determined, difficulties shrunk at his approach and obstacles melted before him. Unprejudiced and pious, he sought for truth in the name and invoking ever the guidance of the great Author of truth. If his theories were not actually deduced from facts, when formed, no test was too severe, and nothing short of a rigid coincidence with fact could satisfy the exacting mind of this wonderful genius. Realising fully the difficulty and importance of the researches before him, once commenced, his perseverance knew no limit, and the fertility of his imagination was utterly inexhaustible.

Such was the man to whom the interests of science at this critical juncture were committed. Having adopted as an hypothesis, the central position of the sun, and the revolution of the earth and planets around this centre, he determined to discover the true nature of the planetary orbits, and find, if possible, some single curve which would explain the orbitual motions of the celestial bodies. To accomplish this difficult enterprise, Kepler wisely determined to confine his efforts and investigations to one single planet, and Mars was selected as the subject for experiment. He commenced by a rigorous comparison between the observed places of the planet, and those given by the best tables which could be computed by the circular theory. Sometimes the predicated and observed places agreed well with each other, and hope whispered that the true theory had been found; but pursuing the planet onward in its sweep around the sun, it would begin to diverge from its theoretic track, its distance would increase, until it became evident that the theory was false, and must be abandoned.

Nothing daunted, the ardent philosopher consoled himself with the thought, that among all possible theories which the mind could frame, one had been stricken from the list, and a diminished number remained for examination. This was a new mode of research; and in case the number of theories was not too great, and the patience of the philosopher sufficiently enduring, a time would come, sooner or later, when success must reward his labours. Thus did Kepler toil on, subjecting one hypothesis after another to the ordeal of rigid experiment, until no less than nineteen had been tested with the utmost severity, and all were rejected. Eight years of incessant labour had been devoted to this examination. He had exhausted every combination of circular motion which the fertility of his imagination could suggest. They had all utterly failed. The charm was ended, and he finally broke away from the fascination of this beautiful curve, which for five thousand years had so bewildered the human mind, and boldly pronounced it impossible to explain the planetary motions with *any* circular hypothesis. This at least was a great negative triumph. If he had not found the curve in which the planets revolved, he had found what it could *not* be; and, released from all future embarrassment from eccentrics and epicycles, he now pursued a lofty and independent train of investigation.

Leaving for ever the circle, the next simplest curve is called the ellipse, an oval figure, which when but little flattened very nearly resembles the circle in form, but enjoys very different properties. All diameters of a circle are equal. The diameters of an ellipse are unequal. The centre of the circle is equally distant from all points on the circumference. No such point exists in the ellipse; but two curious points are found on its longest diameter, possessing the remarkable property of having the sum of the lines joining them with any point of the curve constantly equal to the longest diameter. Each of these points is called a *focus*. This beautiful curve, with its singular properties, had been discovered by the Greek mathematicians; but not remarking its use in nature, it had hitherto been regarded only as an object of amusing speculation. To this curve did Kepler apply, when driven from the circular hypothesis, and again commenced his system of forming hypotheses, and hunting them down, as he termed his scrutinising process. As in the circular hypothesis the sun had at first been located in the centre, so, in commencing the elliptic theory, the centre of the longest diameter was made the centre of motion. Buoyant with hope, the astronomer sets out to follow the planet around its elliptic orbit; but although for a short distance its movements were well represented, it finally broke away from the elliptic track, and bid defiance to the central hypothesis. But Kepler was not in the least disheartened with this first effort. He now shifts the sun to the focus of the ellipse, constructs his orbit, starts once more on the track of the planet, watches it as it sweeps onward around the sun; the elliptic orbit holds it as it moves, farther, and still farther. Half its revolution is performed, and there is no diverging; onward it flies, the goal is won. Triumph crowns the philosopher, *the orbit is found !*

Thus was accomplished one of the most important discoveries which the mind had ever reached. The elliptic orbit of Mars rapidly led to those of the other planets, and to that of the moon, and Kepler proclaimed to the world his first great law, in the following language: "*Planets revolve in elliptic orbits about the sun, which occupies the common focus of all these orbits.*"

This law swept for ever from the heavens and from astronomy those complications which had stood the test of centuries, nay of thousands of years. Their mysterious power was paralysed by this single touch of the enchanter's wand, and they fled from the skies. The circle was as simple and beautiful as ever; but its divine character was gone, and the gods or angels who had so long held their abodes in the planets were exiled from their homes. The dawn of *modern* science broke in beauty on the world.

Kepler having been so signally rewarded by this great discovery, now turned his attention to an investigation of the first importance,—one, indeed, which was indispensably necessary to render his first discovery available. As the planets were known to revolve in ellipses, and as their motion was found by observation to be unequal in different parts of their orbits, it became a matter of the first consequence to ascertain some simple law, regulating the orbitual motion, and by means of which a planet might be readily followed, and its places computed. To detect this law, in whose existence Kepler seems to have entertained the most unwavering faith, a figure was drawn representing the orbit of Mars,—the sun occupying one of the foci of the curve. On the circumference of this curve the places of the planets were marked down as observation had determined them; and here commenced a series of examinations which finally led to the knowledge of the second great law of the planetary motions, which may be thus announced : *If a line be drawn from the centre of the sun to any planet, this line, as it is carried forward by the planet, will sweep over equal areas in equal portions of time.* This law accorded in the most perfect manner with fact, and gave at once the power of following, and, from the mean motion, computing the place of any planet,—a triumph which all the complexity of older systems had failed ever to accomplish.

Any other mind less adventurous than that of Kepler might have been satisfied with these two great discoveries. The precise curves described by the planets, and a law regulating their motions in their orbits, sufficed to render all the phenomena of the heavenly bodies not only explicable, but susceptible of accurate prediction. There seemed nothing more to be added. Kepler did not think so. He conceived the idea that the solar system was not a mere assemblage of isolated planets revolving about a common centre, but a great associated system, in which some common bond of union existed, which, once found, would present the solar system in a new and true light.

This bond he believed existed in some hidden relation between the times occupied by the planets in describing their orbits, and their distances from the sun. In the history of this remarkable research, we are presented with one of the brightest examples of the fruits of perseverance. If some superior power, some spirit from a brighter world, had revealed to the mind of Kepler the actual existence of some relation between the planet's periods and distances, and had proposed to him to discover this hidden law, there would have been a definite object before the astronomer, and to have persevered in the pursuit of this object would have been within the limits of probability, even if a lifetime were exhausted in fruitless efforts. But to excite in his own mind a faith sufficiently strong in the existence of a law of which there existed not the slightest evidence, and to have persevered in its research for seventeen long years of laborious effort, seems almost incredible.

There is an immense difference between the pursuit which resulted in the discovery of the first two laws of Kepler, and the third one. In seeking for the curve described by the planets, it was looking for that which must have an existence ; and in tracing the law of a planet's motion, it was absolutely impossible to follow a planet, or predict its positions, without such a law. But in seeking for a bond of union among the planetary periods and distances, it was a search for that which, it was believed, had no existence, except in the wild imagination of this extraordinary philosopher. The history of mind scarcely furnishes an example in any degree parallel, if we except, perhaps, the heroic fortitude which marked the career of Columbus. Yet even the great Genoese was in possession of solid facts on which to base his reasoning. He saw evidences of the existence of another hemisphere, which the superficial could never realise. Kepler, more bold, more grand, more sublime, dreamed of nothing less than a brotherhood of worlds, a mighty and magnificent scheme of vast revolving orbs. Should success crown his efforts, the most brilliant results would follow. The distance of a single planet from the sun once obtained, and the periodic time of all being known, the distances might then be found for each individual in the entire system, without even directing an instrument to the heavens. Here then was a prize to reach which no time, or pains, or labour could be misapplied. Its return would be a hundredfold.

But where was the prize to be sought? Even admitting that some common bond did bind the circling worlds into one harmonious system, did it exist in some hidden relation between their periods of revolutions, their distances, their magnitudes, their densities? or was it to be sought in some analogy between the distances and periodic times? After long and deliberately pondering this great problem, Kepler decided that the strongest probability suggested that the distances of the planets, and their periods of revolution, would in some way contain the mysterious bond of union. Here then did this daring mind concentrate its energies; and his purpose once fixed, he marches steadily forward in his research with a courage which no defeat could daunt, and a perseverance which knew no limit but success.

Before announcing the final result, let me explain two terms employed in its statement. The *square* of any quantity results by multiplying it by itself. The *cube* comes from multiplying the *square* by the number. The square of a planet's period, or the cube of its distance, are known the moment we know the period and distance, by applying the simple rules of arithmetic. After Kepler had exhausted all simple relations between the periods and distances of the planets, in no degree shaken in his lofty faith, he proceeded to try all possible relations between the squares of the periods and distances, but with as little success. Nothing daunted, he proceeded to investigate the possible relations between the cubes of the periods and distances. Here again he was foiled; no law exhibited itself. He returned ever fresh to the attack, and now commenced a series of trials involving the relations between the simple periods and the squares of the distances. Here a ray of hope broke in upon his dim and darkened path.

No actual relation existed, yet there was a very distant approximation, enough to excite hope. He then tried simple multiples of the periods and the squares of the distances : all in vain. He finally abandoned the simple periods and distances, and rose to an examination of the relations between the squares of these same quantities. Gaining nothing here, he rose still higher, to the cubes of the periods and distances ; no success : until, finally, he tried the proportion existing between the squares of the periods in which the planets perform their revolutions and the cubes of their distances from the sun. Here was the grand secret; but, alas! in making his numerical computations, an error in the work vitiated the results, and with the greatest discovery which the mind ever achieved in his very grasp, the heart-sick and toil-worn philosopher turned away almost in despair from his endless research.

Months rolled round, and yet his mind, with a sort of keen instinct, would recur again and again to this last hypothesis. Guided by some kind angel or spirit, whose sympathy had been touched by the unwearied zeal of the mortal, he returned to his former computations, and with a heaving breast and throbbing heart, he detects the numerical error in his work, and commences anew. The square of Jupiter's period is to the square of Saturn's period as the cube of Jupiter's distance is to some fourth term, which Kepler hoped and prayed might prove to be the cube of Saturn's distance. With trembling hand he sweeps through the maze of figures ; the fourth term is obtained ; he compares it with the cube of Saturn's distance. They are the same! He could scarcely believe his own senses. He feared some demon mocked him. He ran over the work again and again ; he tried the proportion, the square of Jupiter's period to the square of Mars' period as the cube of Jupiter's distance to a fourth term, which he found to be the cube of the distance of

Mars; till finally full conviction burst upon his mind: he had won the goal, the struggle of seventeen, long years was ended, God was vindicated, and the philosopher, in the wild excitement of his glorious triumph, exclaims:

"Nothing holds me. I will indulge my sacred fury! If you forgive me, I rejoice; if you are angry, I can bear it. The die is cast. The book is written, to be read either now, or by posterity, I care not which. It may well wait a century for a reader, since God has waited six thousand years for an observer!"

More than two hundred years have rolled away since Kepler announced his great discoveries. Science has marched forward with swift and resistless energy. The secrets of the universe have been yielded up under the inquisitorial investigations of god-like intellect. The domain of the mind has been extended wider and wider. One planet after another has been added to our system; even the profound abyss which separates us from the fixed stars has been passed, and thousands of rolling suns have been descried, swiftly flying or majestically sweeping through the thronged regions of space. But the laws of Kepler bind them all; satellite and primary, planet and sun, sun and system, all with one accord proclaim, in silent majesty, the triumph of the hero philosopher.

Gregory.

LECTURE IV.

DISCOVERY OF THE GREAT LAWS OF MOTION AND GRAVITATION.

HE remarkable discoveries which had rewarded the researches of Kepler, confirmed in the most perfect manner the doctrines of Copernicus, flowing as they did from his prominent hypothesis, the central position of the sun. Having reached to the true laws of the planetary motions, the whole current of astronomical research was changed. New methods were demanded, and more delicate means of observation must be brought into use before the data could be furnished for new discoveries. Henceforward astronomy could only advance by the aid of kindred sciences. Mathematics, optics, and, above all, mechanical philosophy, were to become the instruments of future conquests.

The philosophy of Aristotle, though very far from deserving it, wielded quite as extensive an influence over the age as did the astronomy of Ptolemy. It appears, indeed, that the followers of Aristotle regarded their master as absolutely infallible, and gave to his doctrines a credence so firm, that even the clearest experiments, the most undeniable evidence of the senses, were sooner to be doubted than the doctrines of the divine Greek. To attack and destroy a system so deeply rooted in the prejudices of the age, required a mind of extraordinary courage and power, a mind deeply imbued with the love of truth, quick in its perceptions, logical in argument, and firm in the hour of trial.

Such a mind was that of the great Florentine philosopher, Galileo Galilei, the senior, friend, and contemporary of Kepler. Indeed, the exigencies of the age seem to have given birth to three men, whose peculiar constitutions fitted them for separate spheres, each of the highest order, each in some measure independent, and yet all combining in the accomplishment of the great scientific revolution. While Tycho, the noble Dane, immured within the narrow limits of his little island, watching from his sentinel-towers the motions of the stars, noting with patient and laborious continuity the revolutions of the sun, moon, and planets, was accumulating the materials which were

Aristotle.

to furnish the keen and inquisitive mind of Kepler with the means of achieving his great triumphs, Galileo, with a giant hand, was shaking to their foundations the philosophical theories of Aristotle, and startling the world with his grand mechanical discoveries. But for the observations of Tycho, Kepler's laws could not have been revealed ; but for the magic tube of Galileo, these laws had been the *ne plus* of astronomical science. Thus do we witness the rare spectacle of three exalted intellects contemporaneously putting forth their diverse talents in the accomplishment of one grand object. The Dane, the German, and the Italian, divided by language and by country, united in the pursuit of science and of truth.

Called to Pisa to discharge the duties of a philosophical teacher, Galileo was not long in detecting the extravagant philosophical errors of Aristotle, which had been implicitly received for more than twenty centuries. He continued to teach the text of his old master ; but it was only to expose its unsound and false philosophy to his wondering and incredulous pupils. A desecration so monstrous could not long escape exposure and punishment. Indeed the Florentine made no secret of his teachings. The Aristotelians made common cause against the young philosophical heretic, and he was warned to desist from his heresy. Galileo gave for answer to his opponents, that he was ready to relinquish his new views the moment they were shown by experiment to be false ; on the other hand, he demanded of them equal candour, and proposed to refer the matter of controversy to the tribunal of experiment.

Galileo's Villa.

Aristotle, in discussing the laws of falling bodies, affirmed the principle, that the velocity acquired by any falling body was in the direct proportion of its weight ; and if two bodies of unequal weight commenced their descent from the same height, at the same moment, the heavier would move as many times swifter than the lighter as its weight exceeded that of the smaller body. Galileo doubted the truth of this principle, and on subjecting it to the test of experiment, he saw instantly that its variation from fact was as wide as it could be. The obvious character of this experiment, its freedom from all chances of deception, and the importance of the principle involved, induced the young philosopher to select it as the test, and to challenge his opponents to a public demonstration of the truth or falsehood of their old system of philosophy. The challenge was accepted. The leaning tower of Pisa presented the most convenient position for the performance of these experiments, on which Galileo so confidently relied for triumphant demonstration of the error of Aristotle; and thither, on the appointed day, the disputants repaired, each party perhaps with equal confidence. It was a great crisis in the history of human knowledge. On the one side stood the assembled wisdom of the universities, revered for age and science, venerable, dignified, united, and commanding.

Galileo.

Around them thronged the multitude, and about them clustered the associations of centuries. On the other, there stood an obscure young man, with no retinue of followers, without reputation, or influence, or station. But his courage was equal to the occasion; confident in the power of truth, his form is erect, and his eye sparkles with excitement.

But the hour of trial arrives. The balls to be employed in the experiments are carefully weighed and scrutinised to detect deception. The parties are satisfied. The one ball is exactly twice the weight of the other. The followers of Aristotle maintain that when the balls are dropped from the top of the tower, the heavy one will reach the ground in exactly half the time employed by the lighter ball. Galileo asserts that the weights of the balls do not affect their velocities, and that the times of descent will be equal; and here the disputants join issue. The balls are conveyed to the summit of the lofty tower. The crowd assembles round the base; the signal is given; the balls are dropped at the same instant; and, swift descending, at the same moment they strike the earth. Again and again is the experiment repeated, with uniform results. Galileo's triumph was complete; not a shadow of doubt remained. But far from receiving, as he had hoped, the warm congratulations of honest conviction, private interest, the loss of place, and the mortification of confessing false teaching, proved too strong for the candour of his adversaries. They clung to their former opinions with the tenacity of despair, and assailed the now proud and haughty Galileo with the bitter feelings of disappointment and hate.

The war was now openly declared, and waged with a fierceness which seems to have excited the mind of the young philosopher to the most extraordinary efforts. Driven from Pisa, by the number and influence of his enemies, no suffering or danger could drive from his mind the great truths which his researches by experiment were constantly revealing. His spirit was unbroken; and, in retiring from the unequal contest, he hurled back defiance into the face of his conquered, though triumphant persecutors.

The mechanical investigations of Galileo, conducted with clearness and precision, soon led to the most important discoveries. He detected the law of falling bodies, and showed that the spaces described were proportional to the squares of the times; that is, if a body fell ten feet in one second of time, it would fall four times as far in two seconds, nine times as far in three seconds, and so on for any number of seconds. He studied with success the subject of the composition of forces, and demonstrated this remarkable proposition, which lies at the very foundation of all modern mechanical philosophy. It may be thus stated. If a body receive an impulse which, singly, would cause it to move thirty feet in a second on the line of the direction of the impulse, and at the same instant another impulse be communicated in a different direction from the first, and which, if acting alone, would cause the body to move on the line of direction of the second impulse forty feet in one second, under the joint action of these two impulses the body will move in a direction easily determined from those of the impulsive forces, and will fly with a velocity of fifty feet in the first second of time.

Such is the universal prevalence of this beautiful proposition,

F

that no falling, flying, or moving body, whether it be the rifle-ball, the cannon shot, or the circling planet, is free from its imperious sway. Strike the knowledge of this great truth from existence, and the magnificent structure which modern science has reared, falls in ruins at a single blow. It is founded in the simple but invariable laws of motion ; and while these endure, this elegant discovery of the Florentine philosopher will remain as a monument to his sagacity and penetration.

Possessed of such rare qualities for philosophic research, so free from prejudice, and withal so candid, we cannot but inquire with interest, how the mind of Galileo stood affected towards the new astronomical doctrines of Copernicus. He had early adopted and taught the Ptolemaic system, and his conversion is so remarkable, and is so characteristic of the man, that it cannot be omitted. A disciple of Copernicus visited the city of Galileo's residence, and delivered several public lectures to crowded audiences, on the new doctrines. Galileo, regarding the whole subject as a species of solemn folly, would not attend. Subsequently, however, in conversing with one who had adopted these new doctrines, the Copernican sustained his views with such a show of reason, that Galileo now regretted that he had heedlessly lost the opportunity of attending the lectures. To make amends, he sought every opportunity to converse with the Copernicans, and remarking that they, like himself, had all once been Ptolemaists, and that from the doctrines of Copernicus no one had ever subse-quently become a follower of the old philosophy, he resolved to examine the subject with the most serious attention. The result may be readily anticipated : the conversion was sudden and thorough, the old astronomy was abandoned, and the new con-vert became the great champion by whose ardour and uncon-querable zeal the strongholds of antiquated systems were to be destroyed, and a new and truthful one founded.

Thus far the career of Galileo in science had been successful and brilliant. He was rapidly rising in reputation and influence, when a fortunate accident revealed to the world the application of a principle in optics, fraught with consequences which it is impossible to estimate. Galileo was informed that Jansen, of Holland, had contrived an instrument possessing the extraordinary property of causing distant objects, viewed through it, to appear as distinctly as when brought near to the eye. The extensive knowledge which Galileo possessed of optics, immediately gave him the command of the important principle on which the new instrument had been constructed. He saw at once the high value of such an instrument in his astronomical researches, and with his own hands commenced its construction.

After incredible pains, he finally succeeded in constructing a telescope, by whose aid the power of the eye was increased thirty fold. It is impossible to conceive the intense interest with which the philosopher directed, for the first time, his wonderful tube to the inspection of the heavens. When we reflect that with the

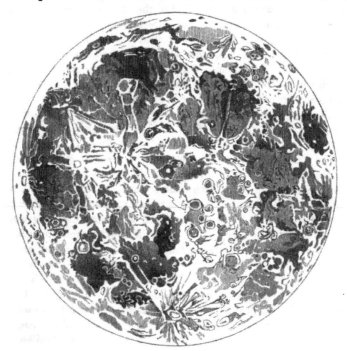

Map of the Moon.

aid of this magical instrument the observer was about to sweep out through space, and to approach the moon, and planets, and stars, to within a distance only one-thirtieth of their actual distance ; that their size was to increase thirty fold, and their distinctness in the same ratio, it is not surprising that these wonders should have excited the most extravagant enthusiasm. Galileo commenced by an examination of the moon. Here he beheld, to his inexpressible delight, the varieties of her surface clearly defined ; her deep cavities, her lofty mountains, her extensive plains, were distinctly revealed to his astonished vision. Having satisfied himself of the reality of these inequalities of the moon's surface, by watching the decreasing shadows of the mountains, as the sun rose higher and higher on the moon, he turned his telescope to an examination of the planets. These objects, which the human eye had never before beheld other than brilliant stars, now appeared round and clear and sharp, like the sun and the moon to unaided vision. On the 8th January, 1610, the telescope was, for the first time, directed to the examination of the planet Jupiter. Its disc was clearly visible, of a pure and silver white, crossed near the centre by a series of dark streaks or belts. Near the planet Galileo remarked three bright stars which were invisible to the naked eye. He carelessly noted their position with reference to the planet, for he believed them to be fixed stars, and of no special interest, except to point out the change in Jupiter's place. On the following night, induced, as he says, by he knew not what cause, he again directed his attention to the same planet. The three bright stars of the preceding evening were still within the field of his telescope, but their positions with reference to each other, were entirely changed, and such was the change, that the orbitual motion of Jupiter could in no way account for it. Astonished and perplexed, the eager astronomer awaits the coming of the following night to resolve this mysterious exhibition. Clouds disappoint his hopes, and he is obliged to curb his impatience. The fourth night was fair, the examination was resumed, and again the bright attendants of Jupiter had changed ; his suspicions were confirmed ; he no longer hesitated, and pronounced these bright stars to be moons, revolving about the great planet as their centre of motion. A few nights perfected the discovery ; the fourth satellite was detected, and this astounding discovery was announced to the world.

No revelation could have been more important or more opportune than that of the satellites of Jupiter. The advocates of the Copernican theory hailed it with intense delight; while the sturdy followers of Ptolemy stoutly maintained the utter absurdity of such pretended discoveries, and urged as a sort of unanswerable argument, that as there were but seven openings in the head, two ears, two eyes, two nostrils, and the mouth, there could be in the heavens but seven planets. The more rational, however, saw the earth, by this discovery, robbed of its pretended dignity. It commanded the attendance of but one moon, while Jupiter received the homage of no less than four bright attendants. The delighted Copernicans saw in Jupiter as a central orb, and in the orderly revolution of his satellites, a miniature of the sun and his planets, hung up in the heavens, and there placed to demonstrate to all coming generations the truth of the new doctrines.

Another discovery soon followed, which, it is said, the sagacity of Copernicus foresaw would sooner or later be revealed to human vision. It had been urged by the Ptolemaists, that in case Venus revolved about the sun, as was asserted by Copernicus, and reflected to us the light of that luminary, then must she imitate exactly the phases of the moon: when on the side opposite to the sun, turning towards us her illuminated hemisphere, she ought to appear round like the moon; while the crescent shape should appear on reaching the point in her revolution which placed her between the sun and the eye of the observer. As these changes were invisible to the naked eye, the objection was urged with a force which no argument could meet. Indeed, it was unanswerable, and in case the telescope should fail to reveal these changes in Venus, the fate of the Copernican theory was for ever sealed.

The position of Venus in her orbit was computed, the crescent phase due to that position determined, the telescope applied, and the eye was greeted with an exquisite miniature of the new moon. There was the planet, and there was the crescent shape long predicted by Copernicus, received by him and his followers as a matter of faith, now become a matter of sight. The doctrines of Copernicus thus received not only confirmation, but, so far as Venus was concerned, a proof so positive that no scepticism could resist. It is not my design to follow the discoveries of the Florentine philosopher among the planetary orbs. These will be resumed hereafter, when we come to examine more particularly the physical constitution of the planets. I have merely adverted to those discoveries, which became specially important in the discussions between the partisans of the old and the new astronomy.

Admitting the doctrines of Copernicus, and uniting to them the great discoveries of Kepler, let us examine the condition of astronomical science, ascertain precisely the point the mind had reached, and the nature of the investigations which next demanded its attention. From the first of Kepler's laws, the figure of the planetary orbits became known, and the magnitude of the ellipse described by any planet was easily determined.   By observing the greatest and least distances of any planet from the sun, the sum of these distances gave the longer axis of the orbit ; and knowing this important line, and the focus, it became a simple matter to construct the entire orbit.   The line joining the planet with the sun, while the planet occupied its shortest or perihelion distance, gave the position of the axis of the orbit in space, and its plane being determined by its inclination to that of the ecliptic, nothing remained to fix in space the figure, magnitude, and position of the planetary orbits.   The next point was to pursue and predict the movements of these revolving bodies. This was readily accomplished.   A series of observations soon revealed the time occupied by any planet in performing one complete revolution in its elliptic orbit.   Knowing thus the periodic time, and the position of the planet in its orbit at any given epoch, the second law of Kepler furnished the key to its future movements ; its velocity in all parts of its orbit became known, and the mind, swift and true, followed the flying world in its rapid flight through space.   It even went further, anticipated its changes, and predicted its positions with a degree of certainty only limited by the accuracy with which the elements of its orbit had been determined.

The third of Kepler's laws, exhibiting the proportion between the periodic times and the mean distances of the planets from the sun, united all these isolated and wandering orbs into one great family.   Their periods of revolution were readily determined by observation, and an accurate determination of the distance from the sun of a single planet in the group, gave at once the distances of all the remaining ones.   The increased accuracy of the means of observation would render more perfect each successive measure of the earth's distance from the sun, and it seemed now that the mind might stop and rest from its arduous toil, that scarcely anything remained to be done.   The solar system was conquered, and the fixed stars defied the utmost efforts of human power.

How widely does this view differ from the true one !   In fact, the true investigation had not even commenced.   A height had indeed been gained, from whence alone the true nature of the next great problem became visible ; and, standing upon this eminence, the mind boldly propounds the following questions :— Why should the orbits of the planets and satellites be ellipses, rather than any other curve ? What power compelled them to pursue their prescribed paths with undeviating accuracy ? What cause produced their accelerated motion when coming round to those parts of their orbits nearer to the sun ? What power held planet and satellite steady in their swift career, producing the most exquisite harmony of motion, and a uniformity of results as steady as the march of time ?

Here I may be asked, Do not such questions border on presumption ? Are not such inquisitorial examinations touching on the domain of God's inscrutable providences, and would it not be wiser to stop and rest satisfied with the answer to all these questions, that God, who built the universe, governs and sustains it by his power and wisdom ? Doubtless this answer is true, and in its truth man humbly finds his highest encouragement to attempt the resolution of the sublime questions already propounded for examination. Let us admit that the divine will produces all motion, speeds the earth in its rapid flight about the sun, guides the planets and their revolving moons, and poises the sun himself in empty space, as the great centre of life and light and heat to his attendant worlds. Is it not reasonable to believe that the will of the Omnipotent is exerted, according to some uniform system, that this system is law, and that this law is within the reach of man ? To encourage this view, the simple laws of motion had been already revealed, and as these must exert a controlling influence in our future examinations, we proceed to unfold them.

First, then, it was discovered that if any body, situated in space and free to move, receive an impulse capable of giving it a velocity of ten feet in the first second of time, or any other velocity, it will move off in the direction of the impulse for ever in a straight line, and with undiminished and unchanged velocity. The intensity of the primitive impulse determines the velocity of the body which receives it, and the one is precisely proportioned to the other. Again, in case a moving body, while pursuing its flight, receives an impulse in a direction different from its primitive one, its new direction and velocity will be determined by the direction and intensity of the new impulse, according to the principle discovered by Galileo, and already explained. Lastly, in every revolving body a disposition is generated to fly from the centre of rotation. The body seems urged by some invisible force from the centre, and if the velocity be sufficiently increased, no matter how strong the bond which unites it to the centre, it will in the end be severed, and the body, freed from its centre, darts away in a straight line tangent to its former circle of revolution. This power, which urges revolving bodies from the centre of motion, is called the *centrifugal* force, and is proportioned to the squares of the velocity of the revolving body. Hence a cord sufficiently strong to hold a heavy ball revolving round a fixed centre, at the rate of fifty feet in a second, would require to have its strength increased four fold, to hold the same ball, if its velocity should be doubled.

These simple laws, derived from a rigorous examination of those moving bodies subject to man's closer scrutiny, extend their sway through the remotest regions of space. Are these laws necessary qualities of matter? Why should a body, darting away under the action of some impulsive force, pursue for ever its undeviating direction, with undiminished velocity? This effect cannot arise from any necessary property or quality of matter. The law might have been different; the direction of the human body might have slowly varied, the velocity might have increased or decreased in any proportion, and yet the flying body, so far as we can understand, have retained all its physical qualities and properties. No: Divine wisdom has selected these simple and beautiful laws from among a multitude, either of which might have been chosen. Stretching forward, therefore, to the examination of the force by which the planets are retained in their orbits, was it not reasonable to expect that some law might be found governing the application of that mysterious power, and in some way proportioned to the mass of the moving body, and to the orbit which it described in wheeling around the sun. That they were held by some invisible power to their centre of motion, was manifest from the fact that the centrifugal force, generated by the rapidity of their revolution, would have hurled them away from the sun, if not opposed and counterpoised by some equivalent power lodged in the great centre of the planetary orbs. Here was an object worthy the highest ambition of the human mind. No matter what might be the nature of this force, whether it should reside in the sun, or in the planet, or in both; whether it should prove to be a property of matter, or the mere uniform manifestation of the Omnipotent will; the discovery of its law of action would give to the mind the power of penetrating the darkest recesses of nature, and of rising to a knowledge of the profoundest secrets of the universe.

Such is the nature of the investigation propounded to the powerful intellect of Newton. This eminent philosopher, justly regarded as the most extraordinary genius that ever lived, neither originated the question which he undertook to discuss,

Sir Isaac Newton.

nor divined the law of force which he proposed to demonstrate. When Kepler had closed the investigations which led to the discovery of his three great laws, his sagacity at once suggested to his mind the existence of some central force, by whose power the planetary movements were controlled. He had watched the moon circling around the earth, he had scrutinised the ocean tide, whose crested wave seemed to rise and follow the movements of the moon, until he boldly announced that some invisible bond, some inscrutable power, united the one to the other. He even reached the conclusion, that this unknown force resided in the moon; that by its power the waters were heaved from their beds, and caused to follow the moon and imitate its motions. Doubtless the solid earth itself felt this mysterious power, and swayed to its influence; but in consequence of the immobility of its particles, its effects had, thus far, escaped detection. Thus once started on the track, Kepler pursued the speculation. He attributed a similar power to the sun, and extended its controlling influence to the planets. He went yet farther, and conjectured that the law of this unknown force was such, that it diminished as the squares of the distances at which it operated increased. That is, if the intensity of the power which it exerted on a planet where the distance was 100,000,000 miles from the sun, be counted as unity, removing the planet to double the distance, or to 200,000,000 miles, the sun's influence over it would be reduced to one-fourth of its former value.

With Kepler this wonderful conjecture always remained without proof. He had placed it on record, and succeeding philosophers had treated it with greater or less seriousness, according to the estimate which they placed upon the sagacity of its author. Even if Kepler himself had attempted the demonstration of this principle, the data were as yet wanting which would have rendered its accomplishment possible. The period intervening between the time of Kepler and Newton had not been left unimproved. Descartes had revealed the law of centrifugal force, and by one of those extraordinary strokes of genius, occurring once in an age, had fastened the irresistible power of analysis upon geometry, which had given to the mind a force and rapidity in the investigation of the figure of curves and curvilinear motion, which had quadrupled its capacity. By repeated efforts, a more accurate knowledge had been obtained of the circumference and diameter of our earth, and through this the distance of the moon from the earth in the successive points of its orbit had been approximated with still greater precision.

With these advantages, Newton gave the energies of his mind to the demonstration of that principle which had existed with Kepler as a mere conjecture.

Before proceeding to develope the train of reasoning pursued by the great English astronomer, permit me first to prepare the way by a simple and perspicuous exhibition of the method employed in determining the diameter of the earth and the distance of the moon ; two elements which figure conspicuously in the demonstration about to be made, and without a knowledge

Birth-place of Newton.

of which it would have been impossible to proceed. We commence with a determination of the diameter of the earth.

If an observer should start from any point on the surface of the earth in the northern hemisphere, and, fixing his eye upon the north pole of the heavens, should travel directly towards that point, all the stars in the north would appear to rise higher above the horizon as he advanced in his journey. The star which occupied the point immediately above his head when he started, would appear gradually to decline towards the south. If it were possible to travel on the same great circle of the earth entirely around its circumference, the zenith star would appear to pursue an opposite route in the heavens, and would return to its primitive position only on the return of the observer to his point of starting. This, however, is not possible. What the observer can accomplish is this. He may travel north until his zenith star shall appear to have moved south by one degree, or the three hundred and sixtieth part of the circumference of the heavens ; then will he have passed over the three hundred and sixtieth part of the circumference of the earth : all these parts are of equal length ; he measures the one over which he has passed, multiplies its value by three hundred and sixty, and the result gives him the circumference of the earth, from which the diameter of the earth is readily deduced by the well-known proportion which exists between these lines. By this simple method, the diameter of the earth being determined, its radius is known, and we are prepared to explain the process by which the moon's distance may be found.

Let us locate, in imagination, two observers at distant points on the same great circle of the earth, each prepared to measure the angular distance at which the moon appears from the zenith point of each station ; but the zenith of any place is the point in which the earth's radius prolonged reaches the heavens ; the angular distance of the moon from the zenith will exhibit precisely the inclination of the visual ray drawn to the moon's centre with the earth's radius drawn to the place of observation ; the zenith distances being observed at each station, the observers knowing that part of the great circle of the earth by which their stations are separated, come together, compare observations, and construct a figure composed of four lines. Two of these are the radii of the earth drawn to the points of observation. These may be laid down under their proper angle, drawing from their extremities two lines, forming, with the radii, angles equal to the moon's measured zenith distances. These represent the visual rays drawn to the moon ; they meet in a point which determines their length, and if the figure be constructed accurately, it will be found that either of these lines is about sixty times longer than the radius of the earth, or the moon's distance is about 240,000 miles.

We now return to the examination of the great question of a central force, and to the discovery of its law of action. Allow me in the outset to explain, with extreme simplicity, the assumed law, whose truth or falsehood it was required to demonstrate. If any force resided in the sun which could resist the centrifugal force of the planets, or in any primary to resist the centrifugal force of the revolving satellite, it was conjectured that this force would decrease in proportion as the square of the distance increased. In other language, if the planets were arranged at the following distances from the sun, the forces exerted upon them would be represented by the second series, thus :—

Distances, 1   2   3   4   5   6 etc.

Forces, 1 $\frac{1}{4}$ $\frac{1}{9}$ $\frac{1}{16}$ $\frac{1}{25}$ $\frac{1}{36}$ etc.

The measure of the intensity of any force of attraction situated at the centre of the earth or sun, is accurately represented by the velocity it is capable of imparting to a falling body in any unit of time. Experiment shows that that power which causes a heavy body to fall to the earth's surface, is capable of impressing upon it a motion of about 16 feet in the first second of time after its fall commences. In case the force diminishes, as we remove the falling body farther from the centre of attraction, the law of diminution would manifest itself in the diminished amount of motion communicated to the falling body.

Now, if Newton could have carried a heavy body upward, above the earth, until he should gain a height above its surface of 4000 miles ; he would then be twice as far from the centre as when at the surface of the earth. Dropping the heavy body, and measuring accurately the distance through which it passes in the first second of time, in case he finds this to be *one-fourth* of 16 feet, the distance fallen through by the same body in the same time, at the distance of 4000 miles from the earth's centre, the result would have confirmed the law which conjecture had assigned as the law of nature. Could he have mounted one unit higher, gaining an altitude of 8000 miles above the earth's surface, or *three* units from the centre, here repeating his experiment, in case the space passed through by the falling body is now *one-ninth* of 16 feet, it would yet further confirm the truth of the conjectured law. Thus, could he have increased his altitude by one unit or radius of the earth after another, repeating his experiment as each new unit was added to his elevation, finding, in every instance, the law of diminution fulfilled by the falling body, all doubt as to the truth of the law would have been removed, and its foundation in nature would have justly flowed from such a series of experiments.

Here, then, is precisely what must be accomplished to demonstrate the assumed law of gravitation. But since these altitudes of 4000 and 8000 miles could not be reached, might not some change in the distance passed over by a heavy falling body be noticed and measured, if removed from a valley to the top of the highest mountain ? Alas ! the increased distance from the centre of the earth gained by ascending the loftiest mountain on its surface, is almost inappreciable, when compared with the entire distance, 4000 miles. Even if the mountain were 10 miles high, the two elevations at which the experiment might then be performed would be 4000 miles and 4010 miles, and the diminished velocity would not be appreciable, even with the most delicate tests, much less could it be relied on to demonstrate the truth or falsehood of a great principle. Here, then, the mind was brought to a full stop; and for a long time it seemed impossible that the philosopher should conquer the difficulties which rose up in his path, and defied his further advance. Finding it impossible to perform any satisfactory experiment on the earth's surface, the daring mind of Newton conceived the idea of employing the moon itself as the falling body, and of testing the truth of his great theory by its fall towards the earth. But could he reach out his hand, grasp the revolving moon, stop it in its orbit, drop it to the earth, and measure its descent in the first second of time? No: this was impossible. The moon could not be arrested in its career; but is this necessary? Is not the moon, in one sense, constantly falling towards the earth? Newton asserted this to be true, and thus did he prove it.

Stand upon the earth, and stretching outward into space 240,000 miles, there let the moon be located, poised, and fixed in space, on a point of its present orbit. There let us suppose it to receive an impulse in a direction perpendicular to the line which joins it with the earth. By the first law of motion, being free to move, it will sweep off in a straight line, tangent to its present orbit, and will pass over a space in the first second of time proportioned to the intensity of the impulse received. Mark that space, and bring the moon back to its primitive position. Now drop it towards the earth, and as it descends freely under the earth's attraction, mark the space through which it falls in the first second of time. This being known, bring back the moon once again to its starting point. Now combine the impulsive force first given with that power which first caused the moon, when left free to move, to fall to the earth. Let them both act at once: the impulse is given, the moon darts off in a straight line, but is instantly seized by the earth's attraction, which drags it from its rectilineal path, and the two contending forces, ever struggling, neither conquering, exercise a divided empire over the moon; onward she moves, obedient to the impulsive force, bent to her orbit by the action of the earth's attraction. Now the amount by which it is deflected in one second of time, from the straight line it would have pursued, is the *amount precisely, by which it falls to the earth.*

If thus far I have been successful, what remains can readily be accomplished. Newton easily computed, from the known velocity of the moon in its orbit, and from the radius of that orbit, the space through which the moon actually fell towards the earth in one second of time. He next computed the space through which a heavy body would fall towards the earth's surface, if removed from the earth to a distance equal to that of the moon. Now in case these two quantities should prove to be exactly equal, the truth of the demonstration would be complete; the moon did fall through the space required by the assumed law, and in this event the law must be the law of nature. For seventeen long years did this incomparable philosopher, rivalling the example of the immortal Kepler, toil on in this most difficult enterprise. He finally reaches the result; the two quantities are found and compared; but alas! the computed distance through which the moon must fall, in case the law of gravitation were true, differed from the observed distance through which it actually fell, by a sixth part of its value. Any less scrupulous, any less philosophic mind, would have been content with this near approximation, and would have announced the discovery to the world. Not so with Newton. Nothing short of the most rigorous accuracy could satisfy his conscientious regard for truth. His manuscripts are laid aside, and the pursuit for the present abandoned.

Months roll by. Occasionally he returns to his computations, runs over the figures, hoping to detect some numerical error; but all is right, and he turns away. At length, while attending a meeting of the Royal Society in London, he learns that Picard had just closed a more accurate measurement of the diameter of the earth. This was one of the important quantities which entered into his investigation. He returns home, and with impatient curiosity spreads before him his old computations; the new value of the earth's diameter is substituted, he dashes onward through the maze of figures, he sees them shaping their values towards the long sought result; the excitement was more than even his great mind could bear; he resigns to a friend; the work is completed, the results compared, they are exactly equal!

The victory is won ! he had seized the golden key which unlocks the mysteries of the universe, and he held it with a giant's grasp ! There never can come another such moment as the one we have described in the history of any mortal. There are no such conquests remaining to be made. Standing upon the giddy height he had gained, Newton's piercing gaze swept forward through coming centuries, and saw the stream of discovery flowing from his newly discovered law, slowly increasing, spreading on the right hand and on the left, growing broader, and deeper, and stronger, encircling in its flow planet after planet, sun after sun, system after system, until the universe of matter was encompassed in its mighty movement. He could not live to accomplish but a small portion of this great work. Rapidly did he extend his theory of gravitation to the planets and their satellites. Each accorded perfectly with the law, and, rising as the inquiry was pursued, he at length announced this grand prevailing law :

Every particle of matter in the universe attracts every other particle of matter with a force or power directly proportioned to the quantity of matter in each, and decreasing as the squares of the distances which separate the particles increase.

Having reached this wonderful generalisation, Newton now propounded this important inquiry. " To determine the nature of the curve which a body would describe in its revolution about a fixed centre, to which it was attracted by a force proportional to the mass of the attracting body, and decreasing with the distance according to the law of gravitation."

His profound knowledge of the higher mathematics, which he had greatly improved, gave to him astonishing facilities for the resolution of this great problem. He hoped and believed that when the expression should be reached which would reveal the nature of the curve sought, that it would be the mathematical language descriptive of the properties of the ellipse. This was the curve in which Kepler had demonstrated that the planets revolved, and a confirmation of the law of gravitation required that the ellipse should be the curve described by the revolving body, on the conditions announced in the problem.

There happens to be a remarkable class of curves, discovered by the Greek mathematicians, called the *conic sections;* thus named because they can all be formed by cutting a cone in certain directions. The figure of a cone with a circle for its base, and converging to a point, is familiar to all. Cut this cone perpendicular to its axis, remove the part cut, and the line on the surface round the cone will be found to be a *circle*. Cut it again, oblique to the axis, then the line of division of the two parts will be an *ellipse*. Cut again so that the knife may pass downward parallel to the slope of the cone, and in this case your section is a *parabola*. Make a last cut parallel to the axis of the cone, and the curve now obtained is the *hyberbola*.

When Newton reached the algebraic expression which, when interpreted, would reveal the properties of the curve sought, and which he had hoped would prove to be the ellipse, he was surprised to find that it did not look familiar to his eye. He examined it closely; it was not the equation of the ellipse, and yet it resembled it in some particulars. What was his astonishment to find, on a complete examination, that the mathematical expression which he had reached, expressing the nature of the curve described by the revolving body, was the general algebraic expression embracing *all the conic sections*. Here is a most wonderful revelation. Is it possible that, under the law of gravitation, the heavenly bodies may revolve in any or either of these curves? Observation responds to the inquiry. The planets were found to revolve in ellipses; the satellites of Jupiter in circles; and those strange, anomalous, out-lawed bodies, the *comets*, whose motions hitherto had defied all investigation, take their place in the new and now perfect system, sweeping round the sun in *parabolic* and *hyperbolic* orbits.

Thus were these four beautiful curves, having a common origin, possessed of certain common properties, yet diverse in character, mingling in close proximity, and gliding imperceptibly into each other, suddenly transferred to the heavens, to become the orbits of countless worlds. For nearly twenty centuries, they had been the objects of curious speculation to the mathematician; henceforward they were to be given up to the hands of the astronomer, the powerful instrument of his future conquests among the planetary and cometary worlds.

The three great laws of Kepler, to which he had risen with such incredible toil and labour, were now found to flow as simple consequences of the law of gravitation. It is impossible to convey the slightest idea, in discussions so devoid of mathematics, of the incredible change which had thus suddenly been wrought in the mode of investigation. I never have closed Newton's

G

investigation, by which he deduces the nature of a curve, described by a body revolving around a fixed centre, under the law of gravitation, bearing with it consequences so simple yet so wonderful, without feelings of the most intense admiration. I can convey no adequate idea of the difference of the methods employed by Kepler and Newton in reaching the three laws of planetary motion. I see Kepler in the condition of one on whom the fates have fixed the task of rolling a huge stone up some rugged mountain side, to its destined level, within a few feet of the summit. He toils on manfully, heaving and struggling, day and night, in storm and in darkness, never quitting his hold, lest he may lose what he has gained. If the ascent be too steep and rocky, he diverges to the right, then to the left, winding his heavy way zigzag up the mountain side. Years glide by, he grows grey in his toil, but he never falters; onward and upward he still heaves the heavy weight; his goal is in sight, he renews his efforts, the last struggle is over; he has finished his task, the goal is won!

Such was Kepler's method of reaching his laws. Now for Newton's. He stands, not at the base of the mountain, with its long, ascending, rocky sides, but on the top. He starts his heavy stone, it rolls of itself over, slowly over, and once again, and falls quietly to its place. Let me not be misunderstood, in this strange comparison, as detracting in the smallest degree from the just fame that is due to Kepler. But for his sublime discoveries, Newton could never have reached the mountain summit on which he so proudly stood. Standing there, he never forgot by whose assistance he had reached the lofty point, and ever recognised, in the most public manner, his deep indebtedness to the immortal Kepler.

A few words with reference to the rigorous application of Kepler's laws in nature will close this discussion. The first law, announcing the revolution of the planets in elliptic orbits, was now made general, and recognised the revolution of the heavenly bodies in *conic sections:* the circle, ellipse, parabola, and hyperbola.

The second law, fixing the equality of the spaces passed in equal times, by the lines joining the planets to the sun as these were carried round in their elliptic orbits, now became applicable to all bodies revolving about a fixed centre, in any curve, and according to any law.

The third law, recognising the proportion between the squares of the periodic times and the cubes of the mean distances of the planets, was extended to the satellites and to the comets; modified slightly in the case of the larger planets, by taking into account their masses or quantities of matter.

Here we close the era of research by observation. The mind has gained its last grand object. The era of physical astronomy dawns; new and wonderful scenes open, and to the contemplation of these we shall soon invite your attention.

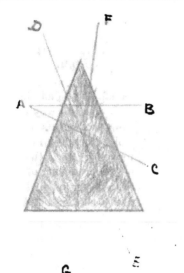

A B = Circle
A C = Ellipse
D E = Parabola
F G = Hyperbola.

LECTURE V.

THE progress of the mind, in its efforts to reach a satisfactory explanation of the movements of the heavenly bodies, previous to the discovery of universal gravitation, had been made independent of any guiding law. The mind had been feeling its way slowly and laboriously, guiding its direction by attentively watching the celestial phenomena, and relying for its success exclusively on the accuracy and number of its observations. Each discovery made was isolated; and although it prepared the way for the succeeding ones, it did not in any sense involve them as necessary consequences. By the discovery of the great law of universal gravitation, a perfect and entire revolution had been made in the science of astronomy. A new department was now added, which, previous to the knowledge of this law, could have no existence. In this branch of astronomy, the process of investigation being inverted, the mind descends from one great law to an examination of its consequences, tracing these in their modified and diversified influences to their final limits. Observation is now employed to verify discovery, and not as the basis on which, and without which, discovery cannot be made.

The era of physical astronomy is, therefore, the great era in the history of the science. It involves the resolution of the most wonderful problems, it calls into use the most refined and powerful mathematical analysis, and demands the application of the most ingenious and delicate instruments in seeking for the data by means of which its theory may be rendered practically applicable to the problems of nature. The mechanical philosopher in his closet may construct his imaginary system. In its centre he locates a sun, containing a certain mass of matter. At any convenient distance from this sun he locates a planet, whose weight he assumes. To this planet he gives an impulse, whose intensity and direction are assumed. The moment these data are fixed, and the impulse given to the imaginary planet, no matter in what kind of an orbit it may dart away, whether circular, elliptical, parabolic, or hyperbolic, the laws of motion and gravitation asserting their empire, the planet is followed by the mathematician, with a certainty and accuracy defying all escape. He assigns its orbit in the heavens, the velocity of its movement, the period of its revolution. In short, in a single line, he writes out its history with perfect accuracy for a million years.

If, now, to this simple system of a great central sun and one solitary planet, the physical astronomer add a third body, a moon, to the planet, he assumes its weight, the intensity and direction of the impulsive force starting it in its career, and now his system becomes more complex. Strike the sun out of existence, fix the planet, and the process of binding the satellite in mathematical fetters is precisely similar to that by which the movements of the planet were prescribed around the sun, before the existence of the satellite. But now, with these three bodies, the train of investigation becomes more intricate and involved. While the planet alone circulated around the sun, such is the undeviating accuracy with which it will for ever pursue its path around the sun, that if it were possible to hang up in space along its route golden rings, whose diameter would just permit the flying planet to pass, millions of revolutions will never mark the slightest change. The rings once passed and then fixed, will mark for ever the pathway of the solitary planet. But the moment a moon is given to this flying world, in that instant its motion is changed ; it is swayed from its original fixed orbit, it no longer passes through the golden rings ; and although the physical astronomer may write out in his analytical symbols the future history of his planet and moon, these expressions are no longer marked with the simplicity which obtained in those which recorded the history of the single planet. While solitary, all changes were effected by the planet in one single revolution, and these were repeated in the same precise order in each successive revolution. Now, with the satellite added, there are changes introduced running through many revolutions, and requiring for their complete compensation vast periods of time. Indeed, the inquiry arises, whether this system of a central sun, with a planet and its satellite revolving about it, can be so constituted that the changes which the planet and its moon mutually produce on each other's movements, may not go on, constantly accumulating in the same direction, until all features of the original orbits of both may be destroyed, both worlds being finally precipitated on the sun, or driven farther and farther from this luminary, until they are lost in infinite space. This inquiry, in a more extended form, will be examined hereafter. We proceed to build up our imaginary system. Thus far, we have regarded our planet and its satellite as mere material heavy points. In case we give to them magnitude and rotation on an axis, the velocity of rotation will determine the figure of the planet and of its satellite. These figures will deviate from the exact spherical form, and this change of figure will sensibly affect the stability of the axis of rotation, and will introduce a series of subordinate movements, each of which must become the subject of research; and to write out the future history of the system, these minute and concealed changes must likewise receive their mathematical expression.

Having thus thoroughly mastered all the phenomena of this system of three bodies, the astronomer now adds another planet, whose mass is assumed, together with the direction and intensity of its primitive impulse. Its orbit is now computed, subject to the greatly predominant influence of the sun, but sensibly affected by the quantity of matter in the old planet and its satellite, which prevents it from forming a fixed and unchangeable orbit in space. Again he is obliged to return to an examination of his first planet and its moon, for these again break away from their previous routes, and in consequence of the action of the second planet, assume new orbits, and are subject to periodical fluctuations, which demand critical examination, and without a knowledge of which no truthful history of the planetary system can be written. To the second planet let us now add several satellites, each of which has its mass assigned, and the direction and intensity of the impulsive force by which they are projected in their orbits. Here, then, is a subordinate system demanding a complete examination. The satellites mutually affect each other's motions, and each is subjected to the influence of the primitive planet and its moon. Again does the physical astronomer review his entire investigation. The addition of these satellites to his second planet has introduced changes in all the previous bodies of the system, which must now be computed, to keep up with the growing complexity. This task is at last accomplished. All the changes are accurately represented. Analysis has mastered the system, and the history of its changes are written out for hundreds and thousands of years.

A third planet with its satellite is now added. This new subordinate system is discussed, and its operation on all the previous planets and satellites computed; and, after incredible pains, the astronomer once more masters the entire group, and follows them all with unerring precision through cycles of changes comprehending thousands or even millions of years.

Thus does the difficulty of grasping the system increase in high ratio by the addition of every new planet and satellite, till finally the last one is placed in its orbit, and the system is complete, so far as planets and satellites are concerned. Through this complicated system now cause thousands of comets to move in eccentric orbits, coming in from every quarter of the heavens. plunging downwards towards the sun, sweeping with incredible velocity around this central luminary, and receding into space to vast distances, either to be lost for ever, or to return after long periods to revisit our system. These wandering bodies must be traced and tracked, their orbits fixed, their periods determined, their influence on the planets and satellites, and that exerted by these on the comets, must be computed and determined; then, and not till then, does the physical astronomer reach to a full knowledge of this now almost infinitely complex system.

In this imaginary problem it will be observed that certain quantities were invariably assumed before the discussion could proceed. The mass of the sun, the mass of each planet and satellite, the intensity and direction of the primitive impulse given to each planet and satellite, these quantities are supposed

to be known. If, now, the astronomer has actually accomplished
the resolution of the imaginary problem, and has obtained analytic expressions which write out and reveal the future history
of his assumed planets and satellites, as they revolve around his
assumed sun, if in these expressions he should substitute the
actual quantities which exist in the solar system for those
assumed, his expressions would then give the history of the solar
system for coming ages, and by reverse action would reveal its
past history with equal certainty.

Before we can, therefore, bring the power of analysis to bear
on the resolution of the grand problem of nature, we must inter-
rogate the heavens, and obtain the absolute weight of our sun, of
each planet, and of every satellite. Next we require the intensity and direction of the impulsive force which projected each
planet and satellite in its orbit, and which would have fixed for ever
the magnitude and position of that orbit, in case no disturbing
causes had operated to modify the action of the primitive impulse.

Having thus attempted to exhibit, at a single view, the general
outlines of the great problem of the solar system, we propose
now to return to the examination of a system composed of three
bodies; and to fix our ideas, we assume the sun, earth, and moon.
In case the earth existed alone, the elliptic orbit described in its
first revolution around the sun would remain unchanged for ever ;
and having pursued it, and marked its changes of velocity in the
different parts of its orbit for a single revolution, this would be
repeated for millions of years. But let us now give to the earth
its satellite, the moon, and setting out from its perihelion, or
nearest distance from the sun, let us endeavour to follow these
two bodies as they sweep together through space, and mark
particularly the effect produced on the moon's orbit by the disturbing influence of the sun. To give to the problem greater
simplicity, let us conceive the plane of the moon's orbit to coincide with the earth's. The law of gravitation, which gives to
every attracting body a power over the attracted one, gravity
increasing as the distance decreases, it will be perceived that
when the earth and moon are nearest to the sun, whatever
influence the sun possesses to embarrass or disturb the motions of
the moon about the earth, will here be exercised with the
greatest effect. But since the sun is exterior to the moon's orbit,
its tendency will be to draw the moon away from the earth, and
cause her to describe around her primary a larger orbit in a

longer period of revolution than would have been employed in case no sun existed, and the moon was given up to the exclusive control of the earth.

Starting the planet on its annual journey, as it recedes from the sun in passing from its nearest to its most remote distance, or from perihelion to aphelion, the moon is gradually removed from the disturbing influence of the sun ; it is subjected more exclusively to the earth's attraction ; its distance from the earth grows less, and the periodic time becomes shorter. These changes continue in the same order until the earth reaches its aphelion. There the moon's orbit is a minimum, and its motion is swiftest. In passing from the aphelion to the perihelion, the earth is constantly approaching the sun ; and as the sun's influence on the moon increases as its distance diminishes, its orbit will now expand by slow degrees, and the periodic time will diminish, until on reaching the perihelion, in case the figure of the earth's orbit remains unchanged, the moon's periodic time will be restored to its primitive value, and all the effects resulting from the elliptic figure of the earth's orbit will have been entirely effaced.

Thus far we have directed our attention exclusively to the changes in the distance of the moon and its periodic time. But the moon's orbit is elliptical, as well as the earth's, and it is manifest that the sun's influence will operate to change not only the magnitude of this orbit, but will in like manner produce a change in the position of the moon's perigee, or nearest point of distance from the earth. If the earth were stationary, and the moon revolved around it, passing between it and the sun, and then coming round so as to be beyond the earth with respect to this luminary, although the moon's orbit would be sensibly affected by the sun's attraction, yet this having exerted itself during one revolution of the moon, all its effects would be repeated in the same order during the next revolution, and the relative positions of the sun and earth remaining the same, the moon would come finally to have a fixed orbit, and its principal lines or axes would never change. But this is not the case of nature. The earth swiftly turning in its orbit, and bearing with it its revolving satellite, by the time the moon has completed a revolution, the sun and earth have entirely changed their relative positions, and the moon cannot reach its perigee, or nearest distance from the earth, at the same point as in the preceding revolution.

By an attentive examination of this problem, it is found that the tendency is to cause the moon to reach its perigee earlier than it would do if not disturbed, and in this way the perigee of a fixed orbit appears to advance to meet the coming moon, and, in the end, to continue advancing until it actually revolves entirely round in a period which observation determines to be about nine years.

It is not my intention to enter into a detailed examination of all the effects resulting from the sun's disturbing power on the moon's motions; neither shall I attempt to exhibit all the effects produced by the moon on the earth. This would require a train of investigation too elaborate and intricate to comport with my present purposes. My object is simply to show that changes must arise from the mutual and reciprocal action of these three bodies, which the theory of gravitation must explain, and the telescope point out, before it be possible to obtain a perfect knowledge of these bodies.

The exact estimation of these changes can never be made until we shall have learned the relative masses of matter contained in the sun, earth, and moon. In other language, we must know how many moons it would require to weigh as much as the earth, and how many earths would form a weight equal to that of the sun.

But is it possible that man, situated upon our planet, 237,000 miles from the moon, and 95,000,000 of miles from the sun, can actually weigh these worlds against each other, and determine their relative masses of matter? Even this has been accomplished, and I shall now proceed to explain how the earth may be weighed against the sun. Dropping a heavy body at the earth's surface, the velocity impressed on it in the first second of time will measure the weight of the earth in one sense. If it were possible to take the same body to the sun, drop it, and measure the velocity acquired by the falling body in the first second of time, the relative distances passed through at the sun and at the earth by the same body in the same time, would show exactly the relative weight of the sun and earth, for their capacity to communicate velocity are exactly proportioned to their masses. Now, although this experiment cannot be performed in the exact terms announced, yet, as we have already shown, the moon is constantly dropping towards the earth, and the earth is as constantly dropping towards the sun. Now, in case we measure the amount by which the moon is deflected from a straight line in one second of time, this measures the intensity of the earth's power. But the amount by which the earth is deflected from a right line by the central power of the sun in one second, is easily measured from a knowledge of its period and the magnitude of its orbit. Executing these calculations, it is found that the sun's effect on the earth is rather more than twice as great as the earth's effect on the moon, and in case these effects were produced at equal distances, then would the sun be shown to contain rather more than twice as much matter as is found in the earth. But the sun produces its effect at a distance four hundred times greater than that at which the earth acts on the moon : hence, as the force diminishes, as the square of the distance increases, a sun acting at twice the distance at which the earth acts, must be four times heavier to produce an equal effect; at three times the distance, it must be nine times heavier, and at four times the distance, sixteen times heavier ; at 400 times the distance, 160,000 times heavier than the earth. Thus do we find that, in case the sun's action on the earth were exactly equal to the earth's action on the moon, in consequence of the great distance at which it operates, its weight would be equal to that of 160,000 earths. But its actual effect is rather more than double that of the earth on the moon, and hence we find it contains rather more than double 160,000 earths, or exactly 354,936 times the quantity of matter contained in the earth.

This enormous mass of the sun is confirmed by an examination of its actual dimensions. An object with an apparent diameter equal to that of the sun, and at a distance of 95,000,000 miles, must have a real diameter of 883,000 miles, a quantity so great that if the sun's centre were placed at the earth's centre, its vast circumference would give ample room for the moon to circulate within its surface, leaving as great a space between the moon's orbit and the sun's surface as now exists between the moon and earth.

It is this immense magnitude of the sun, when compared with the planets and their satellites, which renders the orbits of the planets comparatively unalterable. It is true that these bodies mutually affect each other, but these effects are comparatively slight, and astronomers regard them as *perturbations*, or mere disturbances of the original elliptic motion. Hence we find the magnitude and position of the earth's elliptic orbit remain without any very sensible variation for two or three revolutions; but the slight disturbance experienced at each revolution, constantly accumulating in the same direction, in a long series of years, occasions changes that cannot be lost sight of, and which, by a reflex influence, become in some instances exceedingly important in their practical applications.

As it will be impossible to treat fully the complex subject of perturbation, I will call your attention to a few points about which cluster peculiar interest, in consequence of their great difficulty, and the almost infinite reach of analysis displayed in their successful examination.

I have already explained how it is that the disturbing influence of the sun occasions a constant fluctuation in the periodic time of the moon, accelerating it as the earth moves from perihelion to aphelion, and again retarding it from aphelion to perihelion. If, now, we take a large number of revolutions of the moon, say a thousand, add them all up, and divide by one thousand, we obtain a mean period of revolution, which, in case the earth's orbit remains invariable, will never change, but will be constantly the same for thousands of years. By such an examination during the last century, the mean motion of the moon was obtained with great precision. But on a comparison of eclipses recorded by the Babylonians with each other, it was discovered that the moon in those early ages required a longer time to perform her mean revolution than in modern times. A like comparison of the Babylonian eclipses with those recorded in the middle ages by the Arabian astronomers, confirmed this wonderful discovery, which was yet farther substantiated by comparing the Arabian eclipses with those observed in modern times. It thus became manifest that, to all appearance at least, the moon's mean motion was growing swifter and swifter from century to century; that it was approaching closer and still closer to the earth, and if no limit to this change was ever to be fixed, sooner or later the final catastrophe must come, and the moon be precipitated on the body of the earth, and the system be destroyed.

An effort was made to account for this acceleration of the moon, on the theory of gravitation; but, for a long time, there seemed to be no possibility of rendering a satisfactory explanation of the phenomena, far less of prescribing the limits which should circumscribe the changes. Some, to escape from the difficulty, rejected entirely the ancient eclipses, and boldly cut the knot, by pronouncing the acceleration as impossible, and without any foundation in fact. Others admitted the fact, but finding it impossible to account for it on the hypothesis of gravitation, conceived the idea that the moon was moving in some ethereal fluid, capable of resisting its motion, and producing a diminution in its periodic time of revolution. That *acceleration* should be the effect of *resistance*, may seem to some very strange, but a little reflection will render the subject clear. In case the moon's orbitual motion is resisted, then the centrifugal force, which depends on the velocity, becomes diminished, and the central power of the earth draws the moon closer to itself, decreases the magnitude of its orbit, and, in like manner, reduces the time of accomplishing one revolution about the earth.

Finding no better solution of the mystery, and being obliged to acknowledge the fact that the mean motion of the moon was becoming swifter and swifter, from the action of a resisting medium, there was no escape from the final consequences ; and it was, by some, believed that the elements of decay existed, that the doom of the system was fixed, and although thousands, possibly millions, of years might roll away before the fatal day, yet it must come, slowly, but surely as the march of time. Such was the condition of the problem when Laplace gave the powers of his giant intellect to the resolution of this mysterious subject. The consequences involved gave to it an unspeakable interest, and the world waited with keen anxiety to learn the result of the investigations of this great geometer. Long and difficult was the struggle, slow and laborious the task of devising and tracing out the secret causes of this inscrutable phenomenon. The planets are weighed and poised against the earth, their effects computed on its orbit, the final result of these effects determined, and the reflex influence on the moon's motion computed with the most extraordinary precision. Under the searching examination of Laplace's potent analysis, nature is conquered, the mystery is resolved, the law of gravitation is vindicated ; the system is stable, and shall endure through periods whose limits God alone and not man, shall prescribe.

Follow me in a simple explanation of this most remarkable discovery. It has already been stated, that in case the earth's orbit could remain unchanged, the mean period of the moon as derived from a thousand of its revolutions, would be constant, and would endure without the slightest change for millions of years. But this permanency of the earth's orbit does not exist. Laplace discovered that under the joint action of all the planets, the figure of the earth's orbit was slowly changing; that while its longer axis remained invariable, that its shape was gradually becoming more and more nearly circular. At the end of a vast period, its ellipticity would be destroyed, and the earth would sweep around the sun in an orbit precisely circular. Attaining this limit, a reversed action commences: the elliptic form is resumed by slow degrees; the eccentricity increases from age to age, until, at the end of millions of years, a second limit is reached. The motion is again reversed; the orbit again opens out, approaches its circular form, and thus vibrates backwards and forwards in millions of years, like some mighty pendulum beating the slowly ebbing seconds of eternity!

But do you demand how this change in the figure of the earth's orbit can effect the moon's mean motion? The explanation is easy. Were it possible to seize the earth and hurl it to an infinite distance from the sun, its satellite, now released from the disturbing influence of this great central mass, would yield itself up implicitly to the earth's control. It would be drawn closer to its centre of motion, and its orbit being thus diminished, its periodic time would be shorter, or its motion would be accelerated or made swifter than it now is. This is an exaggerated hypothesis, to render more clear the effect produced by removing the earth farther from the sun. Now, the change from the elliptical to the circular form, which has been progressing for thousands of years, in the earth's orbit, is, so far as it goes, carrying the earth at each revolution a little farther from the sun, releasing, in this way, the moon, by slow degrees, from the disturbing influence of that body; giving to the earth a more exclusive control over the movements of its satellite, and thus increasing the velocity of the moon in its orbit from age to age. But will this acceleration ever reach a limit? Never, until the earth's orbit becomes an exact circle, at the end of millions of years. Then, indeed, does the process change. At every succeeding revolution of the earth in its orbit, its ellipticity returns, its distance from the sun diminishes, the moon is again subjected more and more to the action of the sun, is drawn farther and farther from the earth and its periodic time slowly increases. Thus is acceleration changed into retardation, and at the end of one of these mighty cycles, consisting of millions of years, an exact compensation is effected, and the moon's motion having gone through all its changes, once more resumes its original value.

I can never contemplate this wonderful revolution without feelings of profound admiration. Such is the extreme slowness of this change in the moon's mean motion, that in the period of three thousand years she has got only four of her diameters in advance of the position she would have occupied in case no change whatever had been going on. Here, then, is a cycle of changes, extending backward, to its least limit millions of years, and extending forward to its greatest limit tens of millions of years, detected and measured by man, the existence of whose race on our globe has scarcely been an infinitesimal portion of the vast period required for the full accomplishment of this entire series of changes.

May it not, then, be truly said, that man is in some sense immortal, even here on earth. What is time to him, who embraces changes in swiftly revolving worlds, requiring countless ages for their completion, within the limits of an expression so condensed that it may be written in a single line? Does he not live in the past and in the future, as absolutely as in the present? Indeed, the present is nothing; it is the past and future which make up existence.

In the example of the moon's acceleration just explained, we must not fail to notice a most remarkable fact. It is this: the slow change in the figure of the earth's orbit, occasioned by the joint action of all the planets, and upon which depends the acceleration of the moon's mean motion, is so disguised, that but for its reflex influence on the moon, the probability is it would have escaped detection for thousands of years. The direct effect is almost insensible; but being indirectly propagated to the moon, it is displayed in a greatly exaggerated manner, is in this way detected, and finally, after incredible pains, traced to its origin, and demonstrates in the most beautiful manner the prevalence of the great law of universal gravitation.

Since the general adoption of this law, the human mind has been, in not a few instances, disposed to abandon its universality, and seek for a solution of some intricate problem, by which it was perplexed, in some change or modification of the law; but in no instance has the effort to fly from the law been successful. No matter how long and intricate the examination, how far the mind might be carried from this great law, in the end it must come back and acknowledge its universal empire over our entire system.

It has already been remarked, that one of the effects of the sun's disturbing influence exerted on the moon was to occasion a change in the position of its perigee, causing it to complete an entire revolution in the heavens in about nine years. The theory of gravitation gave a very satisfactory account of this phenomenon generally; but when Sir Isaac Newton undertook the theoretic computation of the rapidity with which the moon's perigee should move, he found, to his astonishment, that no more than one-half of the observed motion of the perigee was obtained from theory. In other language, in case the law of gravitation be true, Newton found that the moon's perigee ought to require eighteen years to perform its revolution in the heavens; while observation showed that the revolution was actually performed in one-half of this period. This great philosopher exhausted all his skill and power in the vain effort to overcome this difficulty. He died, leaving the problem unresolved, bequeathing it to his successors as a research worthy of their utmost efforts.

Astronomers did not fail to recognise the high claims of this investigation. Gravitation was once more endangered. The most elaborate computations were made, and the results obtained by Newton were so invariably verified by each successive computer, that it seemed utterly impossible to avoid the conclusion that they were absolutely accurate, and that the theory of gravitation must be modified in its application to this peculiar phenomenon. At length the problem was taken up by the distinguished astronomer Clairaut. After repeating, in the most accurate manner, the extensive computations of his predecessors, reaching invariably the same results, he finally abandoned the law of gravitation in despair, pronounced it incapable of explaining the phenomenon and undertook to frame a theory which should be in accordance with the facts.

This startling declaration of Clairaut excited the greatest interest. An abandonment of the theory of gravitation was nothing less than returning once more to the original chaos which had reigned in the planetary worlds, and of commencing again the resolution of the great problem which it had long been hoped was entirely within the grasp of the human intellect. In this dilemma, when the physical astronomer had abandoned the law of gravitation in despair, and the legitimate defenders of the theory were mute, an advocate arose where one was least to be expected. Buffon, the eminent naturalist and metaphysician, boldly attacked the new theory of Clairaut, pronounced it impossible, and defended the law of gravitation by a train of general reasoning which the astronomer felt almost disposed to treat with ridicule. What should a naturalist know of such matters? was rather contemptuously asked by the astronomer. It is true he knew but little; yet his attack on Clairaut had the effect of inducing the now irritated astronomer to return to his computations, with a view to overwhelm his adversary. He now determined to rest satisfied with nothing short of absolute perfection. A certain series, which had been reached by every computer, and the value of whose terms had been regarded as decreasing by a certain law, until they finally became inappreciable, from their extreme minuteness, and therefore might, without sensible error, be rejected, was found, on a more careful examination, to undergo a most remarkable change in its character. It was true that the value of its terms did decrease till they became exceedingly small; but so far from becoming absolutely nothing, on reaching a certain value, the decrease became changed into increase; the sum of the series expressing the velocity of the moon's perigee was in this way actually doubled, and Clairaut found, to his inexpressible astonishment, that the investigation which had been commenced with the intention of for ever destroying the universality of the law of gravitation, resulted in his own defeat, and in the perfect and triumphant establishment of this great law.

Thus far, in our examinations of the moon and earth, we have regarded their orbits as lying in the same plane,—an hypothesis which greatly simplifies the complexity of their motions. This, however, is not the case of nature. The moon revolves in an orbit whose plane is inclined under an angle of about four degrees to the plane of the ecliptic. During half of its journey, it lies above the plane of the earth's orbit, while the remaining part of its route is performed below the ecliptic. Thus does the moon, at each revolution, pass through the ecliptic at two points called the *nodes*, which points, being joined by a straight line, give us the intersection of the plane of the moon's orbit with that of the earth. This line of intersection, called the line of the nodes, but for the disturbing influence of external causes, would remain fixed

H

in the heavens. But we know it to be constantly fluctuating, and in the end performing an entire revolution. The exact amount of this change has been made the subject of accurate examination, and the law of its movement has been found to result precisely from the law of gravitation. Not only is the line of intersection of the plane of the moon's orbit with that of the earth constantly changing, but theory as well as observation has ascertained that a series of changes are equally progressing in the angles of inclination of these two planes. The limits are narrow, but the oscillations are unceasing, complicating more and more the relative motions of these two remarkable bodies.

In the physical examination of the revolution of the planetary orbs by the application of the law of gravitation, the general features of the investigation are greatly simplified by the fact that the planets and satellites may be regarded as spherical bodies, and may in general be treated as though their entire mass were condensed into a material heavy point, situated at their centre. While this statement is true in its broader application to the theory of planetary perturbations, or even in the theory of the sun's action on the planets, especially the more distant ones, it is by no means to be admitted, when we come to a critical examination of the figures of the planets, and the influence exerted by these figures on their near satellites.

In case the earth had been created an exact sphere, and had been projected in its orbit without any rotation on an axis, then would its globular figure have remained without sensible change. But as it revolves swiftly on its axis, the laws of motion and gravitation come in to modify the figure of the earth, and to change it from an exact spherical figure to one which is flattened at the poles and protuberant at the equator. Newton's sagacity detected this result as a necessary consequence of the action of gravitation, and he actually computed the figure of the earth from theory, long before any observation or measurement had created a suspicion that its form was other than spherical. The truly wonderful train of consequences flowing from the spheroidal form of the earth gives to this subject a high interest, and demands as close an examination of its principal features as the nature of our investigations will permit.

Give to the earth, then, an exactly spherical form, and a diameter of 8000 miles, with a rotation on an axis once in twenty-four hours, and let us critically examine the consequences. A particle of matter situated on the equator is 4000 miles from the earth's axis, and since it passes over the circumference of a circle whose radius is 4000 miles, it will move with a velocity of about 1000 miles an hour. As we recede from the equator towards the poles, either north or south, the particles revolve at the extremities of radii constantly growing shorter and shorter, until finally, at the exact pole, there is no motion whatever. But in every revolving body, a centrifugal force is generated, a tendency or disposition to fly from the axis of rotation in a plane perpendicular to this axis.

Such is the power of this centrifugal force, that if it were possible to make the earth rotate *seventeen times* in twenty-four hours instead of once, bodies at the equator would be lifted up by the centrifugal force, and the attraction of gravitation would be counterpoised, if not absolutely overcome. The force of gravity exerts its power in directions passing nearly through the centre of the earth, while the centrifugal force is always exerted in a direction perpendicular to the axis of rotation. The consequence is manifest, that these two forces cannot counterpoise each other, except in their action on particles situated on the equator of the revolving body. Let us consider the condition of a particle situated anywhere between the equator and the pole, and free to move under the joint action of these two forces.

In order that such a particle may be held in equilibrium, the two forces must act on the same straight line, and in opposite directions. This is not the case in question, for gravity draws the particle to the centre of the earth, while the centrifugal force urges it from the axis in a plane perpendicular to that axis. The direction of these two forces is inclined under an angle which is nothing at the equator, and increases from the equator to the poles. But the effect produced by the centrifugal force may always be obtained by the joint action of two forces,—the one directed to the centre of the earth, the other tangent to the earth's surface. Substituting these two forces for the centrifugal force, we perceive that the partial force directed towards the earth's centre is destroyed by gravitation, while the tangential force exerts its full power to move the particle towards the equator of the earth.

This being understood, it is manifest, as particles are coming constantly from both poles towards the equator, that a change

H 2

of figure in the earth must be effected. It becomes protuberant at the equator, and is flattened at the poles.

The question now arises, whether there be any limit to this change of figure. In case the velocity of rotation continues undiminished, is there not reason to fear that the earth will grow more and more protuberant at the equator, heaping up the matter higher and higher, till the figure of the earth be destroyed, and its surface rendered uninhabitable ? Theory has answered this important question ; and it has been fully demonstrated that the figure of the earth cannot pass a limit, which it has even now actually attained, and its present form will not change, from the action of the centrifugal force, in millions of years. A condition of equilibrium has been attained, and all further change is at an end. Indeed, if we examine carefully the subject, we may readily perceive, from the nature of the forces, and the conditions of the problem, that such a result might have been anticipated. As the earth grows more protuberant, changing from the spherical form, the particles must be heaved up the side of this elevated ridge which belts the earth around the equatorial regions, and finally the resistance they meet from the elevation they are obliged to overcome, is quite equal to the moving force, and the two destroy each other. This point attained, equilibrium ensues, and further change becomes impossible.

Such is the beautiful order of nature, such the admirable arrangement for stability and perpetuity, everywhere manifested, that the thought constantly comes to the mind that divine wisdom alone could have framed so admirable a system.

But the question may here arise, is this a mere theoretic result ? Has observation confirmed the theoretic figure ? I answer, that observations, the most numerous and diversified, have all united their harmonious testimony to the truth of these beautiful results. In executing exact measures of the degrees of a meridian passing through the poles round the earth, the length of the degree is found to increase from the equator towards the poles, showing that the curvature is more flattened as we recede from the equator. But a more delicate proof is found in the vibrations of the pendulum. A pendulum of a given length will vibrate with a velocity precisely proportioned to the intensity of the force of gravity which operates on it. But the intensity of gravity decreases as the square of the distance from the centre increases, so that it is manifest that the force of gravity is less at the equator than at the poles, in case the surface at the equator is farther from the centre than at the poles, which is the fact asserted by theory.

This being understood, we are prepared to determine the exact figure of the earth, by transporting a pendulum of given length from the equator to different latitudes north and south. The number of vibrations in one hour being accurately counted at the equator, as we recede north or south, will determine with certainty whether we are approaching to, or going farther from, the earth's centre. These experiments have actually been performed, and with the most satisfactory results. The number of vibrations in an hour increases the farther we go north or south, and in a ratio giving the strongest confirmation to the truth of the earth's figure derived from the theoretic investigations; each combining to show that the polar diameter of the earth is but 7898 miles, while the equatorial diameter is 7924 miles, producing a sort of ridge or belt around the equatorial regions, rising about thirteen miles above the general spherical surface described about the polar axis as a diameter.

More than two thousand years have passed away since a discovery was made, showing that the sun's path among the fixed stars was slowly changing. The point at which it crossed the equatorial line, and which for ages had been regarded as fixed, was finally detected to have a slow retrograde motion, producing the *precession of the equinoxes*. The fact was received; but no depth of penetration, no stretch of intellectual vigour, could divine the cause of this inexplicable change. Another fact was revealed about the same time. It was found by attentive examination, that the north pole of the heavens, the point in which the prolongation of the earth's axis pierces the celestial sphere, was actually changing, by slow degrees, its place among the fixed stars. The bright star which, in former ages, had marked the place of the pole, and whose circle of diurnal revolution was scarcely to be perceived, from its smallness, as centuries slowly glided by, was increasing its distance from the pole, gradually describing around it a circle of greater radius. An attentive examination of the stars near the pole soon demonstrated the fact, that it was an actual motion of the pole, and not of the stars in its neighbourhood.

Now incredible as this statement may appear, modern science has traced these phenomena, the revolution of the equinoctial point, and the movement of the north pole of the heavens, to a common origin, and has demonstrated, in the clearest manner, that they are both consequences of the spheroidal figure of the earth, which we have just examined. It is not my design to enter into an elaborate investigation of this wonderful subject, but, in accordance with the plan already announced, I cannot leave you with a mere announcement of a truth so startling, without some effort to explain how this may be. The subject is difficult, but, favoured by your close attention, I do not despair of rendering it approximately intelligible.

Let us conceive the earth's axis to be a solid bar of iron driven through the centre of the earth, coming out at the poles, and extending indefinitely towards the sphere of the fixed stars. Now turn this axis up until it stands perpendicular to the plane of the orbit in which the earth revolves around the sun. Then do the equator and the ecliptic exactly coincide, and if the fixed stars are at a distance nearly infinite, the point in which the earth's axis prolonged pierces the heavens will appear stationary, so far as the revolution of the earth in its orbit is concerned. Now if this iron axle could be grasped by some giant hand, and drawn away from its upright or perpendicular position, the solid earth would turn with it, and the equator, ceasing to coincide with the ecliptic or plane of the earth's orbit, comes to be inclined to it, under an angle precisely equal to the angle through which the axis has been inclined. It is thus seen that no change can be wrought on the position of the axis, that does not involve a corresponding change in the whole earth, and especially in the plane of the equator, which must ever remain perpendicular to the axis in all its positions.

The reverse of this proposition is equally manifest. If the solid earth be seized at the equator, and be turned up or down, the axis will participate in this movement, and its change will exhibit itself in the changed position of the point in which it meets the celestial sphere. One step more, and the difficulty is surmounted. Conceive a flat wheel of wood floating on still water. Through its centre pass an axle which stands perpendicular to the surface of the wheel and water. So long as the wheel floats level, the axle stands erect, but in case the north half of the wheel is tilted down under the water, the south half at the same time rising out of the water, the axis will tilt towards the *north*. Bring the wheel again to its level position. Now plunge the eastern portion of the wheel below the surface. The axis now is tilted towards. the *east*. The experiment is simple, and shows that, in case the successive portions of the wheel be submerged, the axis will always be tilted towards the point which goes under first. To reverse the experiment : in case we take hold of the axle and turn it east, it sinks the eastern half of the wheel below the surface of the water, while the western half is raised out of it ; and then in case we make the upper extremity of the axis follow round the circumference of a circle whose surface is parallel to that of the water, and whose centre is exactly above the centre of the wheel, it will be seen that, as the axle moves round, successive portions of the wheel are submerged, until finally the water line will have divided the wheel into all its successive halves, and will have successively coincided with every possible diameter of the wheel.

Now for the application. The level surface of the water is the evel plane of the earth's orbit, the wheel is the earth's equator, and the axle is the earth's axis of rotation. One half of the equator is constantly submerged below the plane of the ecliptic, the other half rises above it. But the water line, or the intersection of these two planes, the equinoctial line, cannot remain fixed in the same line. A power does seize the equator, and plunge successive halves of it beneath the plane of the ecliptic, changing perpetually the water line, until finally each half in succession, into which all its diameters can be divided, are sunk below the surface or plane of the ecliptic, thus causing the earth's axis to tilt over towards the portions successively submerged, until it finally sweeps entirely round, and comes to resume its first position.

But do you now demand what power seizes the earth's protuberant equator, and tilts it successively towards every point of the compass ? I answer, that the power is lodged in the sun and moon, and it is their combined action which works out these wonderful results. In case the sun and moon were so situated as always to be in the plane of the earth's equator, then they would have no power to change the position of the equator. But we know that they are not in this plane, except when passing through it, and are found sometimes on the north and sometimes on the south side of it. Wherever either of them may be, the nearest half of the redundant matter about the earth's equator will be more forcibly attracted than the remote half, and the equator will be tilted towards the attracting body, and the axis of the earth will follow the movement of the equator to which it is firmly fixed.

Thus does the earth's whole solid mass sway to the motion of the ring of matter heaped up around the equator, delicately and beautifully sensitive to all the changes in the relative places of the sun and moon. Neither the earth nor its axis are ever, for one moment, released from the action of these remote bodies. However slight the effects, however varied in action, oscillating to every point within certain prescribed limits, the stability is preserved, and the final effect is a small retrograde motion of the equinox at the end of every year, and a slight change in the place of the pole of the heavens.

But there is no isolated matter in the universe. Every particle of matter attracts every other particle of matter, and it is impossible for the sun and moon to exert any influence on the equatorial ring of matter which belongs to our globe, without feeling, in their turn, the reaction of this ring on themselves. The remote and ponderous sun may, in consequence of its vast size and distance, escape from any effect capable of being detected by observation. But this is not the case with the moon. Her proximity to the earth, and diminutive mass, render her peculiarly sensitive to the influence of the redundant matter at the earth's equator, and as her attraction tilts the plane of the earth's equator, so does the equatorial ring tilt the plane of the moon's orbit. These effects have been accurately observed and measured, and, strange to relate, their exact values have exhibited the figure which belongs to the earth with far greater precision than can be obtained from measures on its surface. We may even go farther, for such is the intimate relationship between the earth and its attendant satellite, that there is scarcely a question can be asked with reference to the one, that is not answered by the other.

If we demand the weight of the earth when compared with the sun, the moon answers. If the excess of the equatorial diameter of the earth over the polar be required, the moon answers. If the homogeneity of the interior of the solid earth be required, the moon replies. If the thickness of the earth's crust be sought for, question the moon, and the answer comes. If you would know the sun's distance from the earth, ask the moon. If the permanency of the axis of rotation be required, ask the moon and she alone yields a satisfactory reply. Finally, if curiosity leads us to inquire whether the length of the day and night, the revolution of the earth on its axis, be uniform, or whether it may not have changed by a single second in a thousand years, we go to the moon for an answer, and in each and every instance her replies to all these profound and mysterious questions are clear and satisfactory. How wonderful the structure of the universe! How gigantic the power of the human intellect! If all the stars of heaven were struck from existence; if every planet and satellite which the eye and the telescope descry, inside and beyond the earth's orbit, were swept away for ever, and the sun, earth, and moon, alone remained for the study of man, and as evidences of the being and wisdom of God, in the exquisite adjustments or this system, in the reciprocal influences of its three bodies, in their vast cycles of configuration, in their relative masses, magnitudes, distances, motions, and perturbations, there would remain themes sufficient for the exercise of the most exalted genius, and proof of the being of God, so clear and positive, that no sane mind could comprehend it and disbelieve.

Greenwich Observatory.

LECTURE VI.

THE STABILITY OF THE PLANETARY SYSTEM.

HEN, by the application of a single great law, the mind had succeeded in resolving the difficult problems presented by the motions of the earth and its satellite, the moon, it rose to the examination of the higher and more complicated questions of the stability of the entire system of planets, satellites, and comets, which are found to pursue their courses round ⸤the sun. The number of bodies involved in this investigation, their magnitudes and vast periods of revolution, their great distances from the observer, and the exceeding delicacy of the required observations, combined with the high interest which attaches itself to the final result, have united to render this investigation the most wonderful which has ever employed the energies of the human mind.

To comprehend the dignity and importance of this great subject, let us rapidly survey the system, and, moving outward to its known boundaries, mark the number and variety of worlds involved in the investigation. Beginning, then, at the great centre, the grand controlling orb, the sun, we find its magnitude such as greatly to exceed the combined masses of all its attendant planets. Indeed, if these could all be arranged in a straight line on the same side of the sun, so that their joint effect might be

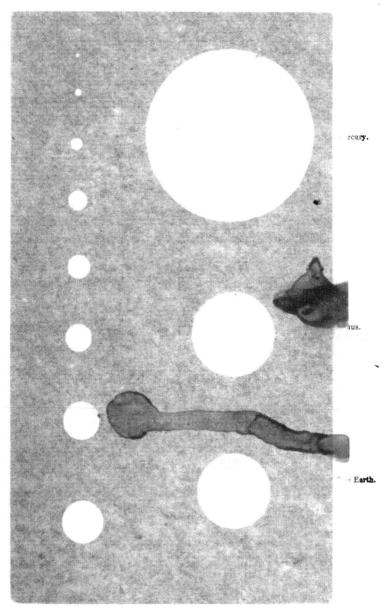

rcury.

us.

Earth.

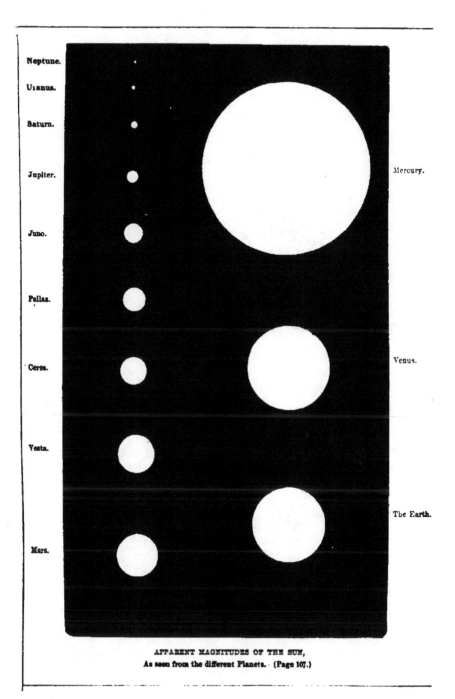

Neptune.

Uranus.

Saturn.

Jupiter.

Mercury.

Juno.

Pallas.

Ceres.

Venus.

Vesta.

The Earth.

Mars.

APPARENT MAGNITUDES OF THE SUN,
As seen from the different Planets.　(Page 107.)

exerted on that body, the centre of gravity of the entire system thus located, would scarcely fall beyond the limits of the sun's surface. At a mean distance of 36,000,000 of miles from the sun we meet the nearest planet, Mercury, revolving in an orbit of considerable eccentricity, and completing its circuit around the sun in a period of about eighty-eight of our days. This world has a diameter of only 3,140 miles, and is the smallest of the old planets. Pursuing our journey, at a distance of 68,000,000 of miles from the sun, we cross the orbit of the planet Venus. Her magnitude is nearly equal to that of the earth. Her diameter is 7,700 miles, and the length of her year is nearly 225 of our days. The next planet we meet is the earth, whose mean distance from the sun is 95,000,000 of miles. The peculiarities which mark its movements, and those of its satellite, have been already discussed. Leaving the earth, and continuing our journey outward, we cross the orbit of Mars, at a mean distance from the sun of 142,000,000 of miles. This planet is 4,100 miles in diameter, and performs its revolution around the sun in about 687 days, in an orbit but little inclined to the plane of the ecliptic. Its features, as we shall see hereafter, are more nearly like those of the earth than any other planet. Beyond the orbit of Mars, and at a mean distance from the sun of about 250,000,000 of miles, we encounter a group of small planets, eight in number, presenting an anomaly in the system, and entirely different from anything elsewhere to be found. These little planets are called asteroids. Their orbits are in general more eccentric, and more inclined to the ecliptic, than those of the other planets ; but the most remarkable fact is this, that their orbits are so nearly equal in size, that when projected on a common plane they are not inclosed, the one within the other, but actually cross each other. We shall return to an examination of these wonderful objects hereafter. At a mean distance of 485,000,000 of miles from the sun, we cross the orbit of Jupiter, the largest and most magnificent of all the planets. His diameter is nearly 90,000 miles. He is attended by four moons, and performs his revolution round the sun in a period of nearly twelve years. Leaving this vast world, and continuing our journey to a distance of 890,000,000 of miles from the sun, we cross the orbit of Saturn, the most wonderful of all the planets. His diameter is 76,068 miles, and he sweeps round the sun in a period of nearly twenty-nine and a half years. He is surrounded by several broad concentric

rings, and is accompanied by no fewer than seven satellites or
moons. The interplanetary spaces we perceive are rapidly
increasing. The orbit of Uranus is crossed at a mean distance
from the sun of 1,800,000,000 of miles. His diameter is 35,000
miles, and his period of revolution amounts to rather more than
eighty-four of our years. He is attended by six moons, and
pursues his journey at a slower rate than any of the interior
planets. Leaving this planet, we reach the known boundary of
the planetary system, at a distance of about 3,000,000,000 of
miles from the sun. Here revolves the last discovered planet,
Neptune, attended by one, probably by two moons, and completing his vast circuit about the sun in a period of one hundred
and sixty-four of our years. His diameter is eight times greater
than the earth's, and he contains an amount of matter sufficient
to form one hundred and twenty-five worlds such as ours.

Here we reach the known limit of the planetary worlds, and
standing at this remote point and looking back towards the sun,
the keenest vision of man could not descry more than one solitary
planet along the line we have traversed. The distance is so great,
that even Saturn and Jupiter are utterly invisible, and the sun
himself has shrunk to be scarcely greater than a fixed star.

There are certain great characteristics which distinguish this
entire scheme of worlds. They are all nearly globular, they all
revolve on axes, their orbits are all nearly circular, they all
revolve in the same direction around the sun, the planes of their
orbits are but slightly inclined to each other, and their moons
follow the same general laws. With a knowledge of these general
facts, it is proposed to trace the reciprocal influences of all these
revolving worlds, and to learn if it be possible, whether this vast
scheme has been so constructed as to endure while time shall
last, or whether the elements of its final dissolution are not contained within itself, either causing the planets, one by one, to
drop into the sun, or to recede from this great centre, released
from its influence, to pursue their lawless orbits through unknown
regions of space.

Before proceeding to the investigation of the great problem of
the stability of the universe, let us examine how far the law of
gravitation extends its influence over the bodies which are united
in the solar system. A broad and distinct line must be drawn
between those phenomena for which gravitation must render a
satisfactory account, and those other phenomena for which it is

in no wise responsible. In the solar system we find, for example, that all the planets revolve in the same direction around the sun, in orbits slightly elliptical, and in planes but little inclined to each other. Neither of these three peculiarities is in any way traceable to the law of gravitation.

Start a planet in its career, and, no matter what be the eccentricity of its orbit, the direction of its movement, or the inclination of the plane in which it pursues its journey, once projected, it falls under the empire of gravitation, and ever after, this law is accountable for all its movements.

We are not, therefore, to regard the remarkable constitution of the solar system as a result of any of the known laws of nature.

If the sun were created, and the planetary worlds formed and placed at the disposal of a being possessed of less than infinite wisdom, and he were required so to locate them in space, and to project them in orbits, such that their revolutions should be eternal, even with the assistance of the known laws of motion and gravitation, this finite being would fail to construct his required system.

Let it be remembered, that each and every one of these bodies exerts an influence upon all the others. There is no isolated object in the system. Planet sways planet, and satellite bends the orbit of satellite, until the primitive curves described, lose the simplicity of their character, and perturbations arise, which may end in absolute destruction. There is no chance work in the construction of our mighty system. Every planet has been weighed and poised, and placed precisely where it should be. If it were possible to drag Jupiter from its orbit, and cause him to change places with the planet Venus, this interchange of orbits would be fatal to the stability of the entire system. In contemplating the delicacy and complexity of the adjustment of the planetary worlds, the mind cannot fail to recognise the fact that, in all this intricate balancing, there is a higher object to be gained than the mere perpetuity of the system.

If stability had been the sole object, it might have been gained by a far simpler arrangement. If God had so constituted matter that the sun might have attracted the planets, while these should exert no influence over each other ; that the planets might have attracted their satellites, while these were free from their reciprocal influences ; then, indeed, a system would have been formed, whose movements would have been eternal, and whose stability would have been independent of the relative positions of the worlds, and the character of their orbits. Give to them but space enough in which to perform their revolutions around the sun, so that no collisions might occur ;—freed from this only danger, every planet, and every satellite, would pursue the same undeviating track throughout the ceaseless ages of eternity.

If this statement be true, it may be demanded why such a system was not adopted. It is impossible for us to assign the reasons which led to the adoption of the present complicated system. Of one thing, however, we are certain : If God designed that in the heavens his glory and his wisdom should be declared, and that in the study of his mighty works, his intelligent creatures should rise higher and higher towards his eternal throne, then, indeed, has the present system been admirably constituted for the accomplishment of this grand design. To have acquired a knowledge of a system constituted of independent planets, free from all mutual perturbations, would have required scarcely no effort to the mind, when compared with that put forth in the investigation of the present complex construction of the planetary system. The mind would have lost the opportunity of achieving its greatest triumphs, while the evidence of infinite wisdom displayed in the arrangement and counterpoising of the present system would have been lost for ever. There is one other thought which here suggests itself with so much force that I cannot turn away from it. We speak of gravitation as some inherent quality or property of matter, as though matter could not exist in case it were deprived of this quality. This is, however, a false idea. Matter might have existed independent of any quality which should cause distant globes to influence each other. This force called *gravitation*, even admitting that it must have an existence, no special law of its action could have forced itself on matter to the exclusion of all other laws. Why does this force diminish as the *square* of the distance at which it operates *increases?* There are almost an infinite number of laws, according to which an attraction might have exerted itself, but there is no one which would have rendered the planets fit abodes for sentient beings, such as now dwell on them, and which would at the same time have guaranteed the perpetuity of the system. Admitting, then, that matter cannot be matter, without exerting some influence on all other matter, (which I am unwilling to admit,) in the selection of the law of the inverse square of the distance, there is the strongest evidence of design.

If we rise above the law of gravitation to the Great Author of nature, and regard the laws of motion and of gravitation as nothing more than the uniform expressions of his will, we perceive at once the impossibility of constructing the universe in such a manner that the sun should attract the planets, without these attracting each other ; or that the planets should attract their satellites without, in turn, being reciprocally influenced by their satellites ; for this would be equivalent to saying that the will of the same Almighty Being should exert itself, and not exert itself, at the same moment, which is impossible. As there is but one God, so there is but one kind of matter, governed by one law, applied by infinite wisdom to the formation of suns and systems without number, crowding the illimitable regions or space, all moving harmoniously, fulfilling their high destiny, and all sustained by the single arm of divine Omnipotence.

We now proceed to an examination of the great question : Is the system of worlds by which we are surrounded, and of which our earth and its moon form a part, so constructed that, under the operation of the known laws of nature, it shall for ever endure, without ever passing certain narrow limits of change, which do not in any way involve its stability ?

It is well known that the planets revolve in elliptical orbits of small eccentricity ; that under the action of the primitive impulse by which they were projected in their orbits, they would have moved off in a straight line with a velocity proportioned to the intensity of the impulse, and which would have endured for ever; but being seized by the central attraction of the sun, at the moment of starting in their career, the joint action of these two forces bends the planet from its straight direction, and causes it to commence a curvilinear path, which carries it round the sun.

The question which first presents itself is this : If the central force lodged in the sun has the power to cause a planet to diverge from the straight line in which, but for this, it would have moved, if it draw it into a curved path, will not this central force, which is ever active, finally overcome, entirely, the impulsive force originally given to the planet, draw it closer and closer to the sun in each successive revolution, in a spiral orbit, until, finally, the planet shall fall into the sun, and be destroyed for ever ?

This question arises independent of the extraneous influence which the planets exert over each other. It refers to a solitary globe revolving around the sun, under the influence of a central force which varies its action as does the law of gravitation. The problem has been submitted to the most rigorous mathematical examination, and a result has been obtained which settles the question in the most absolute manner. The amount by which the central force, in a moment of time, overcomes the effect produced by the primitive impulse, is a quantity *infinitely small*, and of the *second order*. If it were found to be *infinitely small* in each moment of time, then might it accumulate, so that, at the end of a vast period, it might become finite and appreciable. But because it is of the *second order* of *infinitely small quantities*, before it can become an infinitely small quantity of the *first* order, a period equal to infinite ages must roll by, and to make a finite appreciable quantity out of this, an infinite cycle of years must roll round an infinite number of times !

Such is the answer given by analysis to this wonderful question. " Is there no change ? " demands the astronomer. " Yes," answers the all-seeing analysis. " When will it become appreciable ? " asks the astronomer. " At the end of a period infinitely long, repeated an infinite number of times," is the reply.

Having settled this important question, it remains now to examine whether the mutual attractions of the planets on each other may not, in the end, change permanently the form of their orbits, and lead, ultimately, to the destruction of the system. To comprehend more readily the nature of the examination, let us view the points involved in the permanency of our orbit.

Take, for example, our own planet, the earth. It now revolves in an elliptic orbit, whose magnitude is determined by the length of its longer axis, and by its eccentricity. These elements are readily deduced from observation. If it were possible to construct this orbit of some material, like wire, which would permit us to take it up and locate it in space at will, to enable us to give it the position now occupied by the actual orbit of the earth, we must first carry its focus to the sun's centre ; we must then turn its longer axis around this centre as a fixed point, until the nearest vertex of the wire orbit shall fall upon that point of the earth's orbit which is at this time nearest to the sun. Having accomplished this, the axes will coincide in their entire length, and to make the orbits coincident, we must revolve the artificial one around the now common axis, until its plane shall fall upon the actual orbit of the earth.

If, now, change should ever come, in the absolute coincidence of these two orbits, regarding the iron one as fixed and permanent, the orbit of nature may vary from it in any one or all of the following ways : First. The natural orbit, all other things remaining the same, may leave the fixed orbit by a variation of eccentricity ; that is, it may become more or less circular. Second. The planes of the orbits remaining coincident, the curves may separate from each other, in consequence of an angular movement of the longer axis of the natural orbit, by means of which the vertex of the natural curve shall be carried to the right or to the left of the vertex of the fixed one. Third. While these causes are operating to produce change, an increase of deviation may be occasioned by the fact that the two planes may become inclined to each other, thus causing the natural orbit to lie partly above and partly below the fixed one. These, then, are the several ways in which the orbits of the planets may change ; and to settle the question of stability, we must ascertain whether these changes actually exist, and whether any of them, in case they do exist, and are progressing constantly in the same direction, will ever prove fatal to the permanency of the system, finally accomplishing its absolute destruction, or rendering it unfit for the sustenance of that life, which now exists upon the planet.

By a close examination of this great subject, both theoretically and practically, it is found that the system is so constituted, that not a single planet or satellite revolves in an orbit absolutely invariable. Theory demonstrates that such changes must exist, and observation confirms this great truth, by showing that they actually do exist.

Draw, in imagination, a straight line from the sun's centre, through the perihelion; or nearest point to the sun of the earth's orbit, and let it be extended to the outermost limits of the entire system. On this locate the perihelion points of the orbits of all the planets, and in these points fix the planets themselves. They are now all on the same side of the sun, the longer axes of their orbits are in the same direction, and they are all located at their nearest distance from the sun, or in perihelion. The planes of the orbits are inclined to each other under their proper angles, and they all intersect in a common line of nodes passing through the sun's centre. Now give the entire group of planets their

I

primitive impulse, and at the same instant they start in their respective orbits round the sun. Now, in case no perturbations existed, the perihelion points, the inclinations, and the lines of nodes, would remain fixed for ever, and although millions of years might pass away before the planets would again resume their primitive position with reference to each other, yet the time would come when a final restoration would be effected.

At the end of 164 years, Neptune will have completed its revolution round the sun, and will return to its starting point. All the other planets will have performed several revolutions, but each, on reaching the point of departure, will find the perihelion of its orbit changed in position, the inclination altered, and the line of nodes shifted. These changes continue until the longer axes of the orbits, which once coincided, radiate from the sun in all directions. The lines of nodes, once common, now diverge under all angles, the inclinations increasing or decreasing, and even the figures of the orbits undergoing constant mutation; and the grand question arises, whether these changes, no matter how slow, are ever to continue progressing in the same direction, until all the original features of the system shall be effaced, and the possibility of return to the primitive condition destroyed for ever?

Such a problem would seem to be far too deep and complicated ever to be grasped by the human intellect. It is true that no single mind was able to accomplish its complete solution, but the advance made by one has been steadily increased by another, until, finally, not a question remains unanswered. The solution is complete, yielding results of the most wonderful character.

We shall examine this great problem in detail, and commence with the figure of the orbit of any planet, our earth, for example.

The amount of heat received from the sun by the earth depends, other things being the same, on the minor axis of its ecliptic orbit. Any change in the eccentricity operates directly to increase or decrease the shorter axis, and consequently to increase or decrease the mean annual amount of heat received from the sun. Now we know that animal and vegetable life is adjusted in such a way that it requires almost exact uniformity in the mean annual amount of heat which it shall enjoy. An increase or decrease of two or three degrees in temperature would make an entire revolution in the animals and plants belonging to the region experiencing such a change. If, then, it be true that the eccentricity of the earth's orbit is actually changing, under the combined action of the other planets, may this change continue so far as to subvert the order of nature on its surface? This question has been answered in the most satisfactory manner.

It is found that the greater axes of the planetary orbits are subjected to slight and temporary variations, returning, in comparatively short periods, to their primitive values. This important fact guarantees the permanency of the periodic times, so that it becomes possible to deduce, with the utmost precision, the periodic times of the planets, from the mean of a large number of revolutions. That of the earth is now so accurately known, and so absolutely invariable, that we know what it will be a million of years hence, should the system remain as it now is, as perfectly as at the present moment. But neither of these elements secures the stability of the eccentricity, or of the minor axis. Lagrange, however, demonstrated a relation between the masses of the planets, their major axes and eccentricities, such that while the masses remain constant, and the axes invariable, the eccentricity can only vary its value through extremely narrow limits. These limits have been assigned, beyond which the change can never pass, and within these narrow bounds we find the orbits of all the planets slowly vibrating backward and forward, in periods which actually stun the imagination.

This remarkable law for the preservation of the system would not hold in any other organisation. It demands orbits nearly circular, with planes nearly coincident, with periodic times, related as are those of the planets, and the planets themselves located as they actually are. No interchange of orbits is admissible; but, constituted as the system now is, the perpetuity is absolutely certain, so far as the change of eccentricity is concerned.

Let us now examine the changes which affect the position of the major axis in its own plane. The perihelion of every orbit is found to be slowly advancing. Nor is this advance ever to be changed into a retrograde motion. The movement is ever progressive in the same direction, and the perihelion points of all the orbits are slowly sweeping round the sun. That of the earth's orbit accomplishes its revolution in *one hundred and eleven thousand years!* How wonderful the fact, that such discoveries should be made by man, whose entire life is but a minute fraction of these vast periods of time!

Owing to a retrograde motion in the vernal equinox, carrying it around in the opposite direction in 25,868 years, the perihelion and equinox pass each other once in 20,984 years. Knowing their relative positions at this moment, and their rates of motion, it is easy to compute the time of their coincidence. Their last coincidence took place 4,089 years before the Christian era, or about the epoch usually assigned for the creation of man. The effect of the coincidence of the perihelion with the vernal equinox, is to cause an exact equality in the length of spring and summer, compared with autumn and winter. In other language, the sun will occupy exactly half a year in passing from the vernal to the autumnal equinox, and the other half in moving from the autumnal to the vernal equinox.

At present, the line of equinoxes divides the earth's elliptic orbit into two unequal portions. The smaller part is passed over in the fall and winter, causing the earth to be nearer the sun in this season than in summer, and making a difference in the length of the two principal seasons, summer and winter, of some seventeen and a half days. This inequality, which is now in favour of summer, will eventually be destroyed, and the time will come when the earth will be farthest from the sun during the winter, and nearest in the summer. But at the end of a great cycle of more than 20,000 years, all the changes will have been gone through, and, in this respect, a complete compensation and restoration will have been effected.

This epoch of subordinate restoration will find the perihelion of the earth's orbit located in space far distant from the point primitively occupied. Five of these grand revolutions of 20,984 years must roll round before the slow movement of the perihelion shall bring it back to its starting point. 110,000 years will then restore the axis of the earth's orbit, and the equinoctial line, nearly to their relative positions to each other, and to the same region of absolute space occupied at the beginning of this grand cycle.

If now, we direct our attention to the other planets, we find their perihelion points all slowly advancing in the same direction. That of the orbit of Jupiter performs its revolution round the sun in 186,207 years, while the perihelion of Mercury's orbit occupies more than 200,000 years in completing its circuit round the sun. To effect a complete restoration of the planetary orbits to their original position, with reference to their perihelion points, will require a grand compound cycle, amounting to millions of years. Yet the time will come when all the orbits will come again to their primitive positions, to start once more on their ceaseless journeys.

In the changes of the eccentricities, it will be remembered, that the stability of the system was involved. Should these changes be ever progressive, no matter how slowly, a time would finally come when the original figure of the orbit would be destroyed, the planet either falling into the sun, or sweeping away into unknown regions of space. But a limit is assigned, beyond which the change can never pass. Some of the planetary orbits are becoming more circular, others growing more elliptical ; but all have their limits fixed. The earth's orbit, for example, should the present rate of decrease of eccentricity continue, in about half a million of years will become an exact circle. There the progressive motion of the changes stops, and it slowly commences to recover its ellipticity. This is not the case with the motions of the perihelia. Their positions are in no way involved in the well-being of a planet, or in its capacity to sustain the life which exists on its surface ; and since the stability of the system is not endangered by progressive change, it ever continues in the same direction, until the final restoration is effected, by an entire revolution about the sun.

Let us now examine the inclinations of the planetary orbits. Here it is found that there is no guarantee for the stability of the system, provided the angles under which the orbits of the planets are inclined to each other do not remain nearly the same for ever. If changes are found to exist, by which the inclinations are made to increase, without stopping and returning to their primitive condition, then is the perpetuity of the system rendered impossible. Its fair proportions must slowly wear away, the harmony which now prevails be destroyed, and chaos must come again.

Commencing again with the earth, we find that, from the earliest ages, the inclination of the earth's equator to the ecliptic has been decreasing. Since the measure of Eratosthenes, 2078 years ago, the decrease has amounted to about $23'44''$, or about half a second every year. Should the decrease continue, in about 85,000 years the equator and the ecliptic would coincide, and the order of nature would be entirely changed ; perpetual spring would reign throughout the year, and the seasons would be lost for ever. Of this, however, there is no danger. The diminution will reach its limits in a comparatively short time, when the decrease of inclination will change into an increase, and thus slowly rocking backwards and forwards in thousands of years, the seasons shall ever preserve their appointed places, and seed time and harvest shall never fail. These changes of inclination are principally due to the perturbations of Venus, and arising from configurations, will be ultimately entirely compensated.

The angles under which the planetary orbits are inclined to each other are in a constant state of mutation. The orbit of Jupiter at this time forms an angle with the ecliptic of 4731 seconds, and this angle is decreasing at such a rate that, in about 20,000 years, the planes would actually coincide. This would not affect the well-being of the planets or the stability of the system, but should the same change now continue, the angle between the orbits might finally come to fix them even at right angles to each other, and a subversion of the present system would result.

A profound investigation of the problem of the planetary inclinations, accomplished by Lagrange, resulted in the demonstration of a relation between the masses of the planets, the principal axes of their orbits, and the inclinations, such that, although the angles of inclination may vary, the limits are narrow, and they are all found slowly to oscillate about their mean positions, never passing the prescribed limits, and securing, in this particular, the perpetuity of the system.

Here, again, we are presented with the remarkable fact, that whenever mutation involves stability, this mutation is of a compensatory character, always returning upon itself, and, in the long run, correcting its own effects. If all this mighty system was organised by chance, how happens it that the angular motions of the perihelia of the planetary orbits are ever progressive, while the angular motions of the planes of the orbits are vibrating? Design, positive and conspicuous, is written all over the system, in characters from which there is no escape.

We now proceed to an examination of the lines in which the planes of the planetary orbits cut each other, or the lines in which they intersect a fixed plane. These are called the *lines* of *nodes*. They all pass through the sun's centre, and, in case they ever were coincident, they now radiate from a common point in all directions.

Here is an element in no degree involving in its value the stability of the system, and from analogy we already begin to anticipate that its changes, whatever they may be, will probably progress always in the same direction. This is actually the case. The nodes of the planetary orbits are all slowly retrograding on a fixed plane, and in vast periods, amounting to thousands of years, accomplish revolutions, which, in the end, return them to their primitive positions.

Thus are we led to the following results. Of the two elements which fix the magnitude of the planetary orbits, the principal axes, and the eccentricity, the axes remain invariable, while the eccentricity oscillates between narrow and fixed limits. In the long run, therefore, the magnitudes of the orbits are preserved.

Of the three elements which give *position* to the planetary orbits, namely, the place of the perihelion, the lines of nodes, and the inclinations, the two first ever vary in the same direction, and accomplish their restoration at the end of vast periods of revolution, while the inclinations vibrate between narrow and prescribed limits.

One more point, and we close this wonderful investigation. The last question which presents itself is this : May not the periodic times of the planets be so adjusted to each as that the results of certain configurations may be ever repeated without any compensation, and thus, by perpetual accumulation, finally effect a destruction of the system ?

If the periodic times of two neighbouring planets were exact multiples of the same quantity, or if the one was double the other, or in any exact ratio, then the contingency would arise above alluded to, and there would be perturbations which would remain uncompensated. A near approach to this condition of things actually exists in the system, and gave great trouble to geometers. It was found, on comparing observations, that the mean periods of Jupiter and Saturn were not constant ; that one was on the decrease, while the other was on the increase. This discovery seemed to disprove the great demonstration which had fixed as invariable the major axes of the planetary orbits, and guaranteed the stability of the mean motions. It was not until after Laplace had instituted a long and laborious research, that the phenomenon was traced to its true origin, and was found to arise from the near commensurability of the periodic times of Jupiter and Saturn, five of Jupiter's periods being nearly equal to two of Saturn's. In case the equality were exact, it is plain that if the two planets set out from the same straight line drawn from the sun, at the end of a cycle of five of Jupiter's periods, or two of Saturn's, they would be again found in the same relative positions, and whatever effect the one planet had exerted over the other would again be repeated under the same precise circumstances. Hence would arise derangements which would progress in the same direction, and eventually lead to permanent derangement of the system.

But it happens that five of Jupiter's periods are not exactly equal to two of Saturn's, and in this want of equality safety is found. The difference is such, that the point of conjunction of the planets does not fall at the same points of their orbits, but at the end of each cycle is in advance by a few degrees. Thus the conjunction slowly works round the orbits of the planets, and, in the end, the effect produced on one side of the orbit is compensated for on the other, and a mean period of revolution comes out for both planets, which is invariable. In the case of Jupiter and Saturn, the entire compensation is not effected until after a period of nearly a thousand years.

A similar inequality is found to exist between the earth and Venus, with a period much shorter, and producing results much less easily observed. In no instance do we find the periods of any two planets in an exact ratio. They are all incommensurable wth each other, and in this peculiar arrangement we find the stability of the entire system is secured.

So far, then, as the organisation of the great planetary system is concerned, we do not find within itself the elements of its own destruction. Mutation and change are every where found; all is in motion; orbits expanding or contracting, their planes rocking up or down, their perihelia and nodes sweeping in opposite directions round the sun, but the limits of all these changes are fixed; these limits can never be passed, and at the end of a vast period, amounting to many millions of years, the entire range of fluctuation will have been accomplished, the entire system, planets, orbits, inclinations, eccentricities, perihelia, and nodes, will have regained their original values and places, and the great bell of eternity will have then sounded ONE.

Having reached the grand conclusion of the stability of the system of planets, in their reciprocal influences, and that no element of destruction is found in the organisation, we propose next to inquire whether the same features are stamped on the subordinate groups composing the planetary system. As our limits will not permit us to enter into a full examination of all the subordinate groups, we shall confine our remarks to our own earth and its satellite, Jupiter and his satellites, and to Saturn, his rings, and moons. We shall, in this examination, find it practicable to answer, to some extent, the inquiry as to whether either of these systems has received any shock from external causes. We know nothing as to the future, and can, in this particular, only form our conjectures, as to what is to be, from what has been.

We commence our inquiry by an examination of two questions, namely, Is the velocity of rotation of the earth on its axis absolutely invariable ? Has the relation between the earth and moon ever been disturbed by any external cause ? There is nothing so important to the well-being of our planet and its inhabitants, as absolute invariability in the period of its axical rotation. The sidereal day is the great unit of measure for time, and is of the highest consequence in all astronomical investigations. If causes are operating, either to increase or decrease the velocity of rotation, a time will come when the earth will cease to rotate, or else acquire so great a velocity as to destroy its figure, and, in the end, scatter its particles in space.

It is difficult to ascertain from theory a perfectly satisfactory answer to the question of the invariable velocity of rotation of the earth, but Laplace has demonstrated that the length of the day has not varied by the hundredth of one second during the last two thousand years ; that is, the length of the day is neither greater nor less than it was two thousand years ago by the hundredth of a second. The reasoning leading to this remarkable result is simple, and may be readily comprehended by all. Two thousand

Laplace.

years ago, the duration of the moon's period of revolution around the earth was accurately determined, and was expressed in days and parts of a day. The measure of the same period has been accomplished in our own time, and is expressed in days and parts of a day. Now all the causes operating to change the moon's period of revolution are known, and may be applied. When this is done, it is found that the moon's period now and two thousand years ago agree precisely, being accomplished in the same number of days and parts of a day ; which would be impossible, if the unit of measure, the day, had varied ever so slightly.

The extraordinary relation existing between the moon's period and her orbit, and the time occupied in her axical rotation, gives us the opportunity of ascertaining whether our system has received any external shock. These two periods are so accurately adjusted, that in all respects an exact equality exists. The moon ever turns the same hemisphere to the earth, and ever will, unless some external cause should arise to disturb the perfect harmony which now reigns. It is not my purpose to explain why it is that this phenomenon exists. I merely desire to state, that this delicate balancing of periods furnishes an admirable evidence that, for several thousands of years at least, no shock has been received by the earth and its satellite. Steadily have they moved in their orbits, subject only to the influence of causes originating in the constitution of the mighty system of which they constitute a part.

Moving out to a more complex system, we find in the remarkable arrangement of the satellites of Jupiter a delicate test for the action of sudden and extraneous causes. Here we find the periodic times of the satellites so related, that a thousand periods of the first, added to two thousand periods of the third, will be precisely equal to three thousand periods of the second. This delicate balancing of periods would be destroyed by the action of any external shock, such as might be experienced from the collision of a comet sweeping through the system. Thus far, we know that no disturbance has entered, and a knowledge of facts will now pass down to posterity, which will give the means of ascertaining exactly the influence of all disturbing causes which do not form a part of the great system.

The last subordinate group, and the most extraordinary one to which I will at this time direct your attention, is that of Saturn and his rings. Here we find a delicacy of adjustment and equilibrium far exceeding anything yet exhibited in our examinations. This great planet is surrounded certainly by two, probably by three, immense rings, which are formed of solid matter, in all respects like that constituting the central body. These wonderful appendages are nowhere else to be found, throughout the entire solar system, at least with certainty. Their existence has elsewhere been *suspected*, but around Saturn they are seen with a perfection and distinctness which defies all scepticism as to their actual existence. The diameter of the outer ring is no less than 176,000 miles. Its breadth is 21,000 miles, while its thickness does not exceed 100 miles. The inner ring is separated from the outer one by a space of about 1,800 miles, its breadth 34,000 miles, its inner edge being about 20,000 miles from the surface of the planet. Its thickness is the same as that of the outer ring. These extraordinary objects are rotating in the same direction as the planet, and with a velocity so great, that objects on the exterior edge of the outer ring are carried through space with the amazing velocity of nearly 50,000 miles an hour, or nearly fifty times more swiftly than the objects on the earth's equator.

What power of adjustment can secure the stability of these stupendous rings? No solid bond fastens them to the planet; isolated in space, they hold their places, and, revolving with incredible velocity around an imaginary axis, they accompany their planet in its mighty orbit round the sun. Such is the exceeding delicacy with which this system is adjusted, that the slightest external cause once deranging the equilibrium, no readjustment would be effected. The rings would be thrown on the body of the planet, and the system would be destroyed.

To understand the extraordinary character of this system, we must explain a little more fully the three different kinds of equilibrium. The first is called an equilibrium of *instability*, and is exemplified in the effort to balance a rod on the tip of the finger. The slightest deviation from the exact vertical, increases itself constantly, until the equilibrium is destroyed. In case the same rod be balanced on its centre on the finger, it presents an example of an equilibrium of *indifference;* that is, if it be swayed slightly to the one side or the other, there is no tendency to restore itself, or to increase its deviation. It remains indifferent to any change. Take the same rod, and suspend it like a pendulum. Now cause it to deviate from the vertical to the right or left, and it returns of itself to the condition of equilibrium. This is an equilibrium of *stability*. We have already seen that this is the kind of equilibrium which exists in the planetary system. There are constant deviations, but a perpetual effort is making to restore the object to its primitive condition.

Now, in case the rings of Saturn are homogeneous, equally thick, and exactly concentric with the planet, their equilibrium is one of instability. The smallest derangement would find no restorative power, and would even perpetuate and increase itself, until the system were destroyed. For a long time it was believed that the rings were equally thick, and concentric with the planet; but when it was discovered that such features would produce an equilibrium of instability, and that there existed no guarantee for the permanency of this exquisite system, an analytic examination was made, which led to this singular result, namely, To change the equilibrium of instability into one of stability, all that is necessary is to make the ring thicker or denser in some parts than in others, and to cause its centre of position to be without the centre of the planet, and to perform around that centre a revolution in a minute orbit. Finding these conditions analytically, it now became a matter of deep interest to ascertain whether these conditions actually existed in nature. The occasional disappearance of the ring, in consequence of its edge being presented to the eye of the observer, gave an excellent opportunity of determining whether it was of uniform thickness. On these rare occasions, in the most powerful telescopes, the ring remains visible edgewise, and looks like a slender fibre of silver light drawn across the diameter of the planet. In the gradual wasting away of the two extremities of the ring, it has been remarked, that the one remains visible longer than the other. As the ring is swiftly revolving, neither extremity can, in any sense, be regarded as fixed, and hence sometimes the one, sometimes the other, fades first from the sight. An exactly uniform thickness in the ring would render such a phenomenon impossible, and hence we conclude, that the first condition of stability is fulfilled;—the rings are *not* equally thick throughout.

The micrometer was now applied to detect an eccentricity in the central point of the ring. Recent examinations by Struve and Bessel have settled this question in the most satisfactory manner. The centre of the ring does not coincide with that of the planet, and it is actually performing a revolution around the centre of the planet in a minute orbit, thus forming the second delicate condition of equilibrium. The analogy of the great system is unbroken in the subordinate one. For more than two hundred years have these wonderful circles of light whirled in their rapid career under the eye of man, and, freed from all external action, they are so poised, that millions of years shall in no wise affect their beautiful organisation. Their graceful figures and beautiful light shall greet the eyes of the student of the heavens, when ten thousand years shall have rolled away.

Thus do we find that God has built the heavens in wisdom, to declare his glory, and to show forth his handiwork. There are no iron tracks, with bars and bolts, to hold the planets in their orbits. Freely in space they move, ever changing, but never changed; poised and balancing; swaying and swayed; disturbing and disturbed, onward they fly, fulfilling with unerring certainty their mighty cycles. The entire system forms one grand complicated piece of celestial machinery; circle within circle, wheel within wheel, cycle within cycle; revolutions so swift, as to be completed in a few hours; movements so slow, that their mighty periods are only counted by millions of years. Are we to believe that the Divine Architect constructed this admirably adjusted system to wear out, and to fall in ruins, even before one single revolution of its complex scheme of wheels had been performed? No; I see the mighty orbits of the planets slowly rocking to and fro, their figures expanding and contracting, their axes revolving in their vast periods; but stability is there. Every change shall wear away, and after sweeping through the grand cycle of cycles, the whole system shall return to its primitive condition of perfection and beauty.

LECTURE VII.

N the earliest ages of the world, the keen vision of the old astronomers had detected the principal members of the planetary system. Even Mercury, which habitually hovers near the sun, and whose light is almost constantly lost in the superior brilliancy of that luminary, did not escape the eagle glance of the primitive students of the stars. For many thousand years no suspicion arose in the mind, as to the existence of other planets, belonging to the great scheme, and which had remained invisible from their immense distance or their minute dimensions. Indeed the grand investigations which have recently engaged our attention, the mutation of the planetary orbits, their perpetual oscillations and final restoration, the equilibrium of the whole system, had been prosecuted and completed before the mind gave itself seriously to the contemplation of invisible worlds.

The singularly inquisitive genius of Kepler, over whom analogy seems to have ever played the tyrant, in an examination of the interplanetary spaces, finding these to increase with regularity in proceeding outward from the sun, until reaching the space between Mars and Jupiter, which was out of all proportion too great, conceived the idea that an invisible planet revolved in this space, and thus completed the harmony of the system. The space from the orbit of Mercury to that of Venus is 31,000,000 miles ; from the orbit of Venus to that of the earth is 27,000,000 miles ; from the earth's orbit to that of Mars is 50,000,000 miles ; but between the orbit of Mars and that of Jupiter there exists the enormous interval of 359,000,000 miles. The order is again resumed between the orbits of Jupiter and Saturn, and from these slender data Kepler boldly predicted that a time would come when a planet would be found intermediate between the orbits of Mars and Jupiter, whose discovery would establish a regular progression in the interplanetary spaces. For nearly two hundred years this daring speculation was regarded as one of the wild dreams of a great but visionary mind.

Towards the close of the eighteenth century, when the planetary orbs had been studied with great care, and a comparatively accurate knowledge of their perturbations had been reached, certain unexplained irregularities gave rise to the suspicion that the movements of Saturn might be disturbed by the action of an unknown planet revolving in a vast orbit, remote from, and far beyond that of Saturn. These speculations led to no serious results, and it was only by a fortunate accident that, on the 13th of March, 1781, Sir William Herschel noticed a small star of remarkable appearance, which happened to fall in the field of his telescope. On applying a greater magnifying power, the strange star showed

Dr. Herschel.

unequivocal symptoms of increased dimensions. Its position among the neighbouring stars was noticed with care, and by an examination on the following evening, the stranger was found to have sensibly changed position. A few nights sufficed to establish the fact that the newly discovered body was actually a wandering star ; and not for a moment dreaming of the discovery of a new planet, Herschel announced to the world that he had found a remarkable *comet*. Efforts were made to obtain the orbit of the stranger, on the hypothesis, that like those of all the then known comets, it was extremely elongated. Maskelyn and Lexell soon reached the conclusion that no eccentric orbit could possibly represent the motions of the newly-discovered star; and on a close and diligent examination, it was at last discovered to be a primary planet, revolving in an orbit nearly circular, and almost coincident with the plane of the ecliptic. Its motion was progressive, like the other planets, and its vast orbit was only completed at the end of eighty-four of our years. Its distance from the sun was found to be no less than 1,800,000,000 of miles, and its dimensions such that out of it might be formed more than eighty worlds as large as the earth.

This great discovery excited the highest interest in the astronomical world. From the earliest ages, the mighty orbit of Saturn had been regarded as forming the boundary of the vast scheme of planets dependant on the sun. Its slow and majestic motion, its great period and distance, and the wonderful magnificence of its rings and moons, seemed to render it a fitting object to guard the frontiers of the mighty system with which it was associated. But the supremacy of Saturn was now gone for ever, and its sentinel position was usurped by Uranus, whose grand orbit expanded to twice their original dimensions the boundaries of the solar system. Far sweeping in the depths of space, this new world pursued its solemn journey, flinging back the light of its parent orb, steadily obedient to the great law of universal gravitation, which held the old planets true to their changing orbits.

Another unit in the number of interplanetary spaces was thus given, and the law which might possibly regulate the distances of the planets from the sun was sought after with an interest and perseverance which could not long fail of its reward. No exact progression was indeed discovered, but the following remarkable empirical law was detected by Professor Bode:

Write the series　　0　3　6　12　24　48　96　192, &c.

Add to each term　4　4　4　4　4　4　4　4

The sums are　　　4　7　10　16　28　52　100　196

Now if 10 be assumed as the earth's distance from the sun,

the other terms of the series will represent very nearly the distances of the planets, thus ;

4	7	10	16	28	52	100	196
Mercury,	Venus,	Earth,	Mars,	—,	Jupiter,	Saturn,	Uranus.

The fifth term in the series is blank, and falls exactly in the enormous interval which exists between the orbits of Mars and Jupiter, precisely where Kepler had predicted a new planet would be found. As early as 1784, three years after the discovery of Uranus, Baron de Zach, struck with the remarkable law of Bode, even went so far as to compute the probable distance and period of the now generally suspected planet. The impression that a new world would soon be added to the system, grew deeper and stronger in the minds of astronomers, until finally, in 1800, at a meeting held at Lilienthal, by six distinguished observers, the subject was discussed with deep earnestness, and it was finally resolved that the long-suspected, but yet undiscovered world, should be made the object of strict and persevering research. The range of the Zodiac was divided into twenty-four parts, and distributed among an equal number of observers, whose duty it was to scrutinise their particular regions, and detect, if possible, any moving body which might show itself among the fixed stars.

In case it were possible to note down, with perfect precision, the relative places and magnitudes of all the stars in a given region, any subsequent changes which might occur would be easily recognised. In other language, if a daguerreotype picture of any region in the heavens could be made to-night, and if at the end of the year another picture of the same region could be taken, by comparing the number of stars in the one picture with that in the second, in case any one had wandered away from its place, or a stranger had come to occupy a place within the limits of the pictured region, it would be an easy matter to ascertain either the lost star, or the newly arrived stranger. Now, although a daguerreotype picture could not be had, yet, by observation, the exact relative positions of all the visible stars might be mapped out, and a picture formed, which should become the ready means of detecting future changes.

Such was the method of examination adopted by the congress of astronomers assembled at Lilienthal, in 1800. The organisation was made. Baron de Zach was elected president, and

K

Schroeter was chosen perpetual secretary. To those who have paid but little attention to the circumstances under which this strange enterprise was undertaken, nothing can appear more wild and chimerical. To commence a prolonged research for an invisible world, one that no keenness of vision could detect, and which never could be revealed but by telescopic aid, a world of which the magnitude was so small that it would not appear so large as a star of the smallest size visible to the naked eye, and one which must be sought out and detected, not by its planetary disc, but by its wanderings among thousands of stars, which it in all respects resembled, and from which it could be in no wise distinguished but by its motion, seemed like a wasting of time, and an utter throwing away of labour and energy.

Piazzi, of Palermo, in Sicily, was one of the planet-searching association. He had already distinguished himself as an eminent and accurate observer, and had with indefatigable zeal constructed a most extensive catalogue of the relative places of the fixed stars, and thus, in some sense, anticipated a part of the labour that the search for the suspected planet contemplated. Assisted by his own and by preceding catalogues, he entered on the great work with the energy and zeal which distinguished all his great astronomical efforts. On the evening of the first day of the year 1801, this astronomer had his attention attracted by a small star in the constellation of the Bull, which he took to be one recorded in the catalogue of Mayer ; but on examination, it was found not to occupy any place either on Mayer's or his own catalogue. Yet it was so small that it was an easy matter to account for this fact, by its having been overlooked in preceding explorations of the region in which it was found. With intense anxiety, the astronomer awaited the evening of the following night to settle the great question, whether the newly detected star was a fixed or moving body. On the evening of the 2nd of January, he repaired to his observatory, and as soon as the fading twilight permitted, directed the telescope to the exact point in which on the preceding evening his suspicious star had been located. The spot was blank ! But another, which was distant 4' in right ascension, and 3½' in declination, at a spot which on the previous night had certainly been vacant, was now gleaming with the bright little object which on the preceding evening had so earnestly fixed his attention, and for which he was again so anxiously seeking. Night after night he watched its retrograde motion; a motion precisely such as it ought to have, in case it were the long desired planet, until on the 12th it became stationary, and then slowly commenced progressing in the order of the signs. Piazzi was unfortunately taken ill; his observations were suspended, and such was the difficulty of intercommunication, that although he sent intelligence of his discovery to Bode and Orani, associates in the great enterprise, the newly discovered body was already lost in the rays of the sun before it became possible to renew the train of observations by which its orbit might be made known. Piazzi feared to announce the newly-discovered body to be the suspected planet. His observations were few, and he was the only person in the world who had seen it. Bode no sooner received the intelligence of its discovery than he at once pronounced it to be the long-sought planet; and from the scanty materials furnished by Piazzi, Olbers, Burkhart, and Gauss, all computed the elements of its orbit, settled the great fact that it was a superior planet, and that its orbit was included between those of Mars and Jupiter. Some doubt, however, yet rested on the subject, and the disengagement of the planet from the beams of the sun was awaited with the deepest interest.

Several months passed away. Every eye and every telescope was directed to the region in the heavens where the new planet was expected to be found. The most scrutinising search was made for its rediscovery, but without any success. But for the high reputation of Piazzi, his well-known accuracy and honesty, doubts would have arisen as to whether he had not been self-deceived, or was intentionally deceiving others. The subject became of deeper and deeper interest. The world began to sneer at a science which could find a body in the heavens, and then for ever lose it. We must remember that Piazzi had followed it through only about 4° out of 360° of its orbit, and on this narrow basis a research was to be instituted, having for its object the determination of the exact position which the lost planet must occupy. Gauss, then comparatively a young man, and little known as a computer, had conceived a new method of determining the orbits of comets, from a very few and very closely consecutive observations. Here was an admirable opportunity of giving a practical proof of the power of his new method. The long and intricate calculation was finished, the place of the lost planet determined, the telescope was directed.

to the spot, and lo ! the beautiful little orb flashed once more on the eager gaze of the youthful astronomer. For one entire year had the planet been sought in vain, and but for the powerful analysis of Gauss, nothing but years of persevering toil could have wiped away the reproach which rested on astronomy.

A sufficient number of observations were soon made to reveal the orbitual elements of the planet, now named Ceres. It was found in all respects to harmonise in its movements with the older planets, and its orbit filled precisely the blank in the strange empirical law discovered by Bode. The period and distance hypothetically computed from that law sixteen years before, by Baron de Zach, were verified in the most remarkable manner by the actual period and distance of Ceres. Order and beauty now reigned in the planetary system, and a most signal victory had crowned the efforts of astronomical science.

The only remarkable difference between the new planet and the old ones, consisted in its minute size, the great obliquity of its orbit, and the dense atmosphere by which it appears to be surrounded. Its diameter is so small as to render its measure next to impossible ; and the best practical astronomers differ widely in their results. Sir William Herschel makes its diameter only 163 miles, while Schroeter cannot make it less than ten times that quantity. The mean of these two extremes is probably near the truth. No satellites have been found in attendance on this minute planet, although Sir William Herschel suspected the existence of two at one time, a suspicion which subsequent observations have not confirmed.

The beautiful order established in the solar system by the discovery of Ceres was a subject of the highest gratification to the whole astronomical world, and especially to those who had been instrumental in reaching this remarkable result. An opportunity had scarcely presented itself for the expression of delight occasioned by this announcement, before all interested were startled by a declaration from Dr. Olbers, of Bremen, that he had found another planet on the evening of the 28th of March 1802, with a mean distance and periodic time almost identical with those of Ceres. This discovery broke through all the analogies of the solar system, and presented the wonderful anomaly of two planets revolving in such close proximity, that their orbits, projected on the plane of the ecliptic, actually intersected each other.

The new planet was called *Pallas*, and is of a magnitude about equal to that of Ceres. Its orbit is greatly inclined to the plane of the ecliptic, and its eccentricity is very considerable. The existence of these small planets, in such close proximity, for a long while perplexed astronomers. At length Olbers suggested that these minute bodies might be the fragments of a great world, rent asunder by some internal convulsion of sufficient power to produce the terrific result, but of a nature entirely beyond the boundary of conjecture.

Extraordinary as this hypothesis may appear, the results to which it led are not less remarkable. If a world of large size had been actually burst into fragments, it is easy to perceive that these fragments, all darting away in the orbits due to their impulsive forces, would start from the same point, and hence would return at different intervals indeed, but would all again pass through the point of space occupied by the parent orb when the convulsion occurred. Having found two of these fragmentary worlds, the point of intersection of their orbits would indicate the region through which the other fragments might be expected to pass, and in which they might possibly be discovered. So reasonable did the views of Olbers appear, that his suggestions were immediately acted upon by himself and several distinguished observers, and on the 2nd of September, 1804, Mr. Harding, of Lilienthal, while scrutinising the very region indicated by Olbers, detected a star of the eighth magnitude, which seemed to be a stranger, and was soon recognised to be another small planet, fully agreeing, in all its essential characteristics, with the theory of Olbers. The new world was named Juno, and is remarkable for the eccentricity of its orbit. Its diameter has not been well determined, owing to its minute size. This discovery gave to the theory of Olbers the air of reality ; and finding the nodes of the three fragments to lie in the opposite constellations Cetus and Virgo, he prosecuted his researches in those regions with redoubled energy and zeal.

His efforts were not long without their reward. On the 29th of March, 1807, he detected the fourth of his fragments in the constellation Virgo, and very near the point through which he had, for four years, been waiting to see it pass. This was a most wonderful discovery, and almost fixed the stamp of truth upon the most extraordinary theory which had ever been promulgated. This new asteroid was named Vesta, and, for nearly forty years, the examinations which were conducted revealed no new fragment, and it began to be regarded as positively ascertained, that all the small bodies revolving in this region had been revealed to the eye.

But on the 8th of December, 1845, Mr. Hencke announced to the world the discovery of another asteroid, which was named Astrea.

Mr. J. R. Hind.

Before two years had rolled round, the same indefatigable observer discovered a sixth member in this wonderful group, which was called Hebe. His success induced other observers to undertake a similar examination, and, in a very short time, the researches of Mr. Hind, of London, were rewarded by the discovery of a seventh and eighth asteroid, which were named Iris and Flora.

Thus have we no less than eight of these minute worlds, revolving in orbits so nearly equal, that for weeks and months these miniature orbs may sweep along in space, almost within hail of each other.* Let us now return to an examination of the hypothesis of Olbers, that these are the fragments of a world of large size, which once occupied an orbit intermediate between those of Mars and Jupiter.

* _Thirteen_ of these small planets have been discovered; four of them by Mr. Hind, at Mr. Bishop's Observatory, Regent's-park. On April 12, 1849, Signor de Gasparis, of the Observatory of Naples, whilst comparing Steinheil's Star Map for hour XII with the heavens, discovered a planet: its appearance at this time was that of a small star of the 9th or 10th magnitude. M. de Gasparis referred the naming of his new Planet to M. Capocci, who called it _Hygeia_. On May 11, 1850, Dr. Annibal de Gasparis, assistant at the Royal Observatory at Naples, discovered a small planet, and which he named _Parthenope_. On September 2, 1850, Mr. Hind discovered another, making the third, which he has named _Victoria_. The orbits of both these planets are situated between those of Mars and Jupiter. While these sheets were passing through the press, the newspapers announced that Mr. Hind has discovered another new planet, the fourth, in the constellation Scorpio, about 8° north of the ecliptic, and forming at the time an equilateral triangle with the stars Scorpii and Libra. It is stated to be of a pale bluish colour, and that its light is about equal to that of a star of the ninth magnitude.—ED.

If any internal convulsion could burst a world and separate its fragments, it is readily seen that the fragments of largest mass would move in orbits more nearly coincident with that of the original planet, while the smaller fragments would revolve in orbits greatly inclined to the primitive one. This condition is wonderfully fulfilled among the asteroids. The larger planets, Ceres and Vesta, revolve in orbits with small inclinations to the ecliptic, while the smaller objects are, in some instances, found to move in planes with very great inclinations. The force necessary to burst a planet, and to give to its fragments certain orbits, has been computed by Lagrange, and he finds that, in case any fragment is projected with an initial velocity one hundred and twenty-one times greater than that of a cannon ball, it would become a *direct* comet, with a parabolic orbit, while a primitive velocity, one hundred and fifty-six times greater than that of a cannon-ball, would cause the fragments to revolve with a *retrograde* motion in the curve of a parabola. Any less powerful force would cause the fragment to revolve in ellipses ; and it is probable that the force which operated to produce the asteroids was not more than twenty or thirty times greater than that of a cannon-ball. Although the theory of Olbers has received new accessions of strength from the discovery of every new asteroid, it would be wrong to regard it as one of the demonstrated truths of astronomy. In the meantime, powerful efforts are making to scour the heavens, and a method of observation has been proffered to the Academy of Sciences, of Paris, by which all the visible fragments may be discovered within a period of four years. Should this plan, which contemplates a division of the heavens among different astronomers, be adopted, volunteers have already presented themselves, and the most interesting results may be anticipated.

From this curious branch of astronomical inquiry, we turn to one of still deeper interest. In the examinations for new planets, thus far, the telescope has been the sole instrument of research. Conjectures based upon analogical reasoning, it is true, guided the instrumental examinations ; but the mind had never dared to rise to the effort of reasoning its way analytically to the exact position of an unknown body. It has been reserved for our own day to produce the most remarkable and the boldest theorising which has ever marked the career of astronomical science. I refer to the analytic effort to trace out the orbit, define the distance, and weigh the mass of an unknown planet as far beyond the extremest known planet as it is from the sun.

I am fully aware of the difficulties by which I am surrounded, when I invite your attention to this complex and intricate subject; and I know how utterly impossible it is, in a popular effort, to do any kind of justice to the intricate and involved reasoning of the great geometers, who have not only rendered themselves, but the age in which we live, illustrious by their efforts to resolve this, the grandest problem which has ever been presented for human genius. Trusting to your close attention, I shall attempt to exhibit some faint outline of the train of reasoning and the kind of research employed in rescuing an unknown world from the viewless regions of space in which it has been tracing its orbit for ages, commensurate with the existence of the great system of orbs of which it constitutes a part.

After the discovery of the planet Uranus, by Sir William Herschel, geometers were not long in fitting it with an orbit which represented in the outset, with accuracy, its early movements. With this orbit, it became possible to trace its career backwards, and to define its position among the fixed stars for fifty or one hundred years previous to the date of its discovery. This was actually done, with the hope of finding that the place of the planet had been observed and recorded by some astronomer, who ranked it among the fixed stars. This hope was not disappointed. The planet, believed to be a fixed star, had been seen and observed no less than nineteen different times, by four different observers, through a period running back nearly one hundred years previous to the discovery of its planetary character by Herschel. These remote observations were of the greatest value as data for the determination of the elements of its elliptic orbit, and for the computation of the mean places, which might serve to predict its position in coming years.

A distinguished astronomer, M. Bouvard, of the Paris Academy of Sciences, about thirty years ago, undertook the analytic investigation of the movements of Uranus, and a computation of exact tables. He was met, however, by difficulties which, in the state of knowledge as it then existed, with reference to this planet, were absolutely insurmountable. He found it quite impossible to obtain any orbit which would pass through the places of the planet determined after its discovery, and through those positions which had been fixed previous to that epoch. In this dilemma, it became necessary to reject the old observations as less reliable than the new ones, and the learned computer leaves the problem for posterity to resolve, carefully abstaining from any absolute decision in the case.

His orbit, based upon the new or modern observations, and his tables being computed, it was hoped that the theoretic places of the planet would thereafter coincide with the observed places, and that all discrepancies which might not be fairly chargeable to errors of observation, would be removed. In this expectation, however, the astronomical world was disappointed; and while the tables of Bouvard failed absolutely to represent the ancient observations, in a few years they were but little more truthful in giving the positions actually filled by the planet under the telescope. The discrepancies between the theoretic and actual places of the planet began to attract attention many years since. As early as 1838, Mr. Airy, Astronomer Royal of England, on a comparison of his own observations with the tables, found that the planet was out of its computed track, by a distance as great as the moon's distance from the earth, and that it was actually describing an orbit greater than that pointed out by theory. It seemed that this remote body was breaking away from the sun's control, or that it was operated upon by some unknown body deep sunk in space, and which thus far had escaped the scrutinising gaze of man.

These deviations became so palpable as to attract general attention, and various conjectures were made with reference to their probable cause. Some were disposed to regard the law of gravitation as somewhat relaxed in its rigorous application to this remote body; others thought the deviation attributable to the action of some large comet, which might sway the planet from its course; while a third set of philosophers conjectured the existence of a large satellite revolving about Uranus, and from whose attraction the planet was caused to swerve from the computed orbit. These conjectures were not sustained by any show of reasoning, and were of no scientific value.

Such was the condition of the problem when it was undertaken by a young French astronomer, not quite unknown to fame in his own country, but comparatively at the beginning of his scientific career. The friend of Arago, Leverrier's Cometary Investigations, and more especially his researches of the motions of Mercury, had gained for him the confidence of this distinguished *savant*, and Arago urged on his young associate the importance of the great problem presented in the perturbations of

Uranus, and induced him to abandon other investigations, and
concentrate all the energies of his genius on this profound and
complex investigation.

The extraordinary powers of Leverrier as a mathematical
astronomer had been so successfully displayed in his researches
of the motions of Mercury, that it deserves a passing notice.
The old tables of this planet, Leverrier believed to be defective.
He therefore set about a thorough examination of its entire
theory, and after a rigid scrutiny, deduced a new set of tables,
from which the places of the planet might be predicted with
greater precision.

The transit of Mercury across the sun's disc, which occurred
on the 8th day of May, 1845, presented an admirable opportunity
to test the truth of the new theory of the young astronomer.

Leverrier.

Most unhappily for his
hopes, all observations in
Paris were rendered im-
possible by the clouds,
which covered the heavens
during the entire day on
which the transit took
place. While the com-
puter was sadly disap-
pointed, I was more fortunate, for a pure and
transparent atmosphere
favoured this, the first
astronomical observation
I ever made. A slight
reference to this occur-
rence may be pardoned.
For three years I had
been toiling to complete
a most difficult and la-
borious enterprise, the
erection of an astronomical observatory of the first class, in a
country where none had ever existed. Amid difficulties and
perplexities which none can ever know, the work had moved on;
and at length I had the high satisfaction of seeing mounted one
of the largest and most perfect instruments in the world. I had
arranged and adjusted its complex machinery, had computed the
exact point on the sun's disc where this planet ought to make its

first contact, had determined the instant of contact by the old tables, and by the new ones of Leverrier, and, with feelings which must be experienced to be realised, five minutes before the computed time of contact, I took my post at the telescope to watch the coming of the expected planet. After waiting what seemed almost an age, I called to my friend how much time was yet to pass, and I found but one single minute out of five had rolled heavily away. The watch was again resumed. Long and patiently did I hold my place, but again was forced to call out, How speeds the time? and was answered that there was yet wanting *two minutes* of the computed time of contact. With steadfast eye, and a throbbing heart, the vigil was resumed, and after waiting what seemed an age, I caught the dark break which the black body of the planet made on the bright disc of the sun. "Now!" I exclaimed; and, within *sixteen seconds* of the computed time, did the planet touch the solar disc, at the precise point at which theory had indicated the first contact would occur.

The planet was followed across the disc of the sun, round, and sharp, and black; and every observation confirmed the superior accuracy of the new tables of Leverrier. While the old tables were out fully a minute and a half, in the various contacts, those of Leverrier were in error by only about sixteen seconds as a mean.

The great success of this investigation encouraged the young astronomer to accept the difficult task which Arago proposed for his accomplishment, and he earnestly set about preparing the way for a full discussion of the grand problem of the perturbations of Uranus. The importance of the subject demanded the greatest caution, and having determined to rely solely on his own efforts, he at once rejected all that had been previously done, and commenced the problem at the very beginning. New analytic theories were formed, elaborate investigations of the planets Jupiter and Saturn, as disturbing bodies, were made, and an entire clearing up of all possible causes of disturbance in the known bodies of the system was laboriously and successfully accomplished, and the indefatigable mathematician finally reached a point where he could say, "Here are residual perturbations which are not to be accounted for by any known existing body, and their explanation is to be sought beyond the present ascertained limits of the solar system."

As early as the 10th of November, 1845, M. Leverrier

presented a memoir to the Royal Academy of Sciences in Paris, in which he determined the exact perturbations of Jupiter and Saturn on Uranus. This was followed by a memoir, read before the academy on the first of June, 1846, in which he demonstrated that it was impossible to render an exact account of the perturbations of Uranus in any other way than by admitting the existence of a *new planet* exterior to the orbit of Uranus, and whose heliocentric longitude he fixed at 325° on the 1st of January, 1847. On the 30th of August, 1846, a third memoir was presented to the Academy, in which the elements of the orbit of the unknown planet were fixed, together with its mass and actual position, with greater accuracy, giving, on the 1st of January, 1847, 326° 32′ for its heliocentric longitude. Finally, on the 5th of October, 1847, a fourth memoir was read, relative to the determination of the plane of the orbit of the constructive planet.

It is quite impossible to convey, in popular form, the least idea of the profound analytic reasoning employed by M. Leverrier in this wonderful investigation. None but the rarest genius would have dared to reach out 1,800,000,000 miles into unknown regions of space, to *feel* for a planet which had displaced Uranus by an amount only about equal to four times the apparent diameter of the planet Jupiter, as seen with the naked eye ;—a quantity so small that no eye, however keen and piercing, without telescopic aid, could ever have detected it. Yet, from this minute basis was the magnificent superstructure to be reared which should eventually direct the telescope to the place of a new and distant world. To many minds, the resolution of such a problem may appear utterly beyond the powers of human genius, and without one ray of light to illumine the midnight darkness which surrounds it to them, they are disposed to reject the entire subject. An attentive examination of the following train of reasoning may at least demonstrate that the problem is not quite so hopeless as it would at first appear.

It was not necessary to extend researches to all quarters of the heavens indifferently, in an effort to find the unknown body. All the planets revolve in planes nearly coincident with the plane of the earth's orbit, and more especially do the distant ones. Jupiter and Saturn and Uranus revolve in orbits but little inclined to the plane of the ecliptic. Hence, it was fair to conjecture that the new planet, should it ever be found, would not violate this general law, and a search for it was properly limited to a narrow belt near the plane of the earth's orbit.

The limits of research were thus brought down to a narrow zone, sweeping around the entire heavens, indeed, but insignificant in extent, when compared with the whole celestial sphere. .

The next point of examination was the probable distance of the unknown planet. Here, again, analogy came to the aid of Leverrier. The empirical law of Bode, already explained, showed that the remote planets increased their distances by a very simple law. Saturn was twice as remote as Jupiter; Uranus was at double the distance of Saturn, and it was fair to conclude that the unknown planet would be about twice as far from the sun as Uranus. As a first approximation, then, its distance was fixed at about 3,600,000,000 miles from the sun. Kepler's law, regulating the ratio between the distances and periods of the planets, gave at once the time of revolution of the new planet, in case its distance had been correctly assumed. In the next place, it was fair to conclude that the orbit of the new planet, like those of Jupiter, Saturn, and Uranus, would not differ greatly from a circle. These conjectures were, in some degree, confirmed by a very simple train of reasoning with reference to the distance of the disturbing body. If it revolved in an orbit very near to that of Uranus, then its effect on Uranus ought to be excessive, when compared with its influence on Saturn, which was found not to be the case. Again, if it revolved in an orbit very far beyond the limit assigned above, its effect on Uranus· and Saturn would be very nearly the same, which was not verified by examination.

Having thus roughly fixed limits for the orbit of the unknown body, the work of the mathematician now commences, having for its grand object the determination of the true places of the planet sought at some given epoch, and such an orbit as will represent the perturbations of Uranus in the most perfect manner. To exhibit, in some faint degree, the difficulty of this investigation, let us conceive that up to the first of January, 1800, the solar system had consisted only of the known bodies,— the sun, planets, satellites, and comets. The orbits of all the planets are accurately ascertained, and their reciprocal influences computed and known. The outermost planet, Uranus, revolves in its vast orbit obedient to the great law of gravitation, acknowledging the predominant influence of the sun, and swaying more or less to the action of the nearest planets, Saturn and Jupiter. Its predicted and observed places coincide, and its movement is followed with confidence and exactitude.

With a full and perfect knowledge of the orbit of Uranus, let a new planet be created and projected in a vast orbit, exterior to, and remote from, the orbit of Uranus. The new body thus added to the system would instantly derange the motions of Uranus, causing it to recede from the sun, and increasing its mean period of revolution. In this case, the total effect of the new planet on Uranus would be perturbation; and it would not be quite impossible, even for one not skilled in the higher mathematics, to see how the action of the newly created planet on the movements of the old one might actually reveal approximately the position of the disturbing body. It is manifest, that when the two planets are in conjunction, or on a right line passing through them and the sun, that at this configuration the new planet would exert its greatest power to drag the old one outward from the sun; and if it could be found at what point of its orbit the old planet actually receded to its greatest distance from the sun, in the same direction, nearly, must the disturbing body have been situated at that time. In this way, we perceive, one place of the new planet might be approximately found; and from its periodic time it would be possible to trace it backward or forward in its orbit, for the present supposed to be circular.

The problem here presented is certainly sufficiently difficult; yet its complexity is very far from being equal to that presented in nature, and with which the French geometer found himself obliged to grapple. Although unknown, the new planet did exist, and for ages had silently pursued its unknown orbit round the sun. Its influence on Uranus had been ever active; and when the observations on Uranus were made, and its places determined, from which its elliptic elements were to be derived, these very places were in part dependent on the action of the invisible disturber, and hence a portion of its influence would be darkly concealed in the orbit of Uranus; and to divide the entire effect of the new planet on the old one correctly between the disguised portion, and that producing real perturbation, was attended with the greatest difficulty, and could only be reached by adopting certain positive hypotheses. Surrounded by all these difficulties, Leverrier worked on, and with consummate art so constructed his analytical machinery as to meet and master every difficulty; and he finally announced to the world the figure of the orbit of his imaginary planet, its distance, period of revolution, and even the mass of matter it contained.

These important communications were made to the French Academy of Sciences on the 31st of August, 1846. On the 18th of the following month, M. Leverrier wrote to his friend, M. Galle, of Berlin, requesting him to direct his telescope to that point in the heavens which his computations had revealed as the one occupied by the constructive planet. The request was readily complied with, and on the very first evening of examination, a star of the eighth magnitude was discovered, which was evidently a stranger in that region, as it was not found on an accurate map of the heavens, including all stars of that magnitude. The following evening was awaited with the deepest interest, to decide, by the actual motion of the suspected star, whether indeed it was the planet so wonderfully revealed by the analysis of Leverrier. The night came on. Again was the telescope directed to the star in question, when lo ! it had moved from its former place, in a direction and with a velocity almost precisely accordant with the theory of the French geometer ! The triumph was perfect ; the planet was actually found. The news of its discovery flew in every direction, and filled the world with astonishment and admiration.

The exceeding accuracy with which its place had been predicted, coming within less than *one degree* of its actual position, gave to M. Leverrier the highest confidence in the perfection of his analysis, and filled with astonishment the oldest and most learned astronomers. If scepticism had existed with reference to the possible solution of so complex a problem,—if the theory of Leverrier had been regarded as a beautiful speculation, ingenious and plausible, but vain in its practical application,—the actual discovery of the planet silenced all cavil, and put to flight every lingering doubt.

As if anything were wanting to give a more positive character to the computations of Leverrier, it was now found that a young English mathematician, Mr. Adams, of Cambridge, had actually accomplished the resolution of precisely the same problem, and had reached results almost identical with those of the French geometer. This astonishing coincidence on the part of two computers unknown to each other, each starting from the same data, pursuing independent trains of reasoning, and arriving at the same results, confirmed, as it would seem, in the fullest manner, the accuracy of the resolution which had been obtained.

On learning that Leverrier had communicated to the Academy of Sciences, in August, 1846, his final results, I wrote immediately, requesting the computed place of his planet, with such directions as would best guide me in a search, which I desired to make for it with the great refracting telescope of the Cincinnati Observatory. But before my letter reached its destination, the planet had been found, and the news of its discovery soon reached the United States. It was almost impossible for me to credit the statement, and I was almost disposed to believe that the prediction of the planet's position had been mistaken for its actual discovery. With these conflicting doubts, I waited for the coming of night with a degree of anxiety and excitement which may be readily imagined. I had no star-chart to guide me in my search for the planet; I had no meridian instrument with which to detect it by its motion; but I was not without hope that the power of our great telescope might be sufficient to select at once the planet from among the fixed stars, by the magnitude of its *disc*.

As soon as the twilight disappeared, the instrument was directed to the point in the heavens where the planet had been found. I took my place at the *finder*, or small telescope attached to the larger one, and my assistant was seated at the great instrument.

On placing my eye to the finder, four stars of the eighth magnitude occupied its field. One of them was brought into the field of the large telescope, and critically examined by my assistant, and rejected. A second star was in like manner examined, and rejected. A third star, rather smaller and whiter than either of the others, was now brought to the centre of the field of the great telescope, when my assistant exclaimed, "There it is! there is the planet! with a disc as round, bright, and beautiful as that of Jupiter!" There, indeed, was the planet, throwing its light back to us from the enormous distance of more than 3,000,000,000 of miles, and yet so clear and distinct, that in a few minutes its diameter was measured, and its magnitude computed.

It is not my intention to follow, critically, the history of this wonderful discovery; yet there are some facts so remarkable, that it would be wrong to pass them in silence. From the moment the planet was detected in Berlin, it has been observed by all the best instruments in the world, with a view of ascertaining how accurately theory had assigned the elements of its orbit. In consequence of its very slow motion, it became a matter of the utmost importance to obtain, if possible, some remote observation made by an astronomer who might have entered the place of the planet in his catalogue as a fixed star. Mr. Adams led the way in the computation of the elements of the orbit of the new planet from actual observation, and was followed by many other computers,—among them Mr. S. C. Walker, then of the Washington Observatory, United States.

Having obtained an approximate orbit, Mr. Walker computed backwards the places of the new planet for more than fifty years

Lalande.

and then examined the late catalogues, in the hope of finding its place on some of them as a fixed star. Among recent catalogues there was no success; but in an examination of Lalande's Catalogue, he found an observation on a star of the eighth magnitude, made May 10th, 1795, which was so near the place which his computation assigned the planet at the same date, that he was led to suspect that this star might indeed prove to be the new planet. In case his conjecture were true, on turning the telescope to the place occupied by the star, it would be found *blank*, as its planetary motion would have removed it very far from the place which it occupied more than fifty years before. The experiment having been made, no star could be found; and strong evidence was thus presented that Mr. Walker had actually found an observation of the new planet, giving its position in 1795; but in consequence of the great discrepancy between

L

the period of M. Leverrier and that which would result from a reliance on this observation of the new planet Neptune, Mr. Walker's discovery was at first received with great hesitation. A greater doubt was thrown over the matter from the fact that Lalande had marked the observation as uncertain; and it was only by reference to the original manuscripts preserved in the Royal Observatory of Paris that the doubts could be removed.

The discovery of Mr. Walker was subsequently made by Mr. Petersen, of Altona; and the results of these astronomers reached Paris on the same day. A committee was at once appointed to examine the original manuscript of Lalande, when a most remarkable discovery was made. This astronomer had observed a star of the eighth magnitude on the evening of the 8th of May, 1795; and on the evening of the 10th, not finding the star as laid down, but observing one of the same magnitude very near the former place, he rejects the observation of the 8th of May as inaccurate, and enters the observation of the 10th, marking it doubtful.

On close examination, this star proves to be the planet Neptune; and by this discovery we are placed in possession of observations which render it possible to determine the elliptic elements of the new planet with great precision. These differ so greatly from those announced by Leverrier and Adams previous to the discovery, that Professor Pierce, of Cambridge, Massachusetts, pronounces it impossible so to extend fairly the limits of Leverrier's analysis as to embrace the planet Neptune; and that, although its mass, as determined from the elongation of its satellite, renders it possible to account for all the perturbations of Uranus by its action, in the most surprising manner, yet, in the opinion of Professor Pierce, it is not the planet to which geometrical analysis directed the telescope. Leverrier rejects absolutely the result reached by the American geometer, and claims Neptune to be the planet of his theory, in the strictest and most legitimate sense.

Time and observation will settle the differences of these distinguished geometers; and truth being the grand object of all research, its discovery will be hailed with equal enthusiasm by both of the disputants. In any event, the profound analytic research of Leverrier is an ever-during monument to his genius, and his name is for ever associated with the most wonderful discovery that ever marked the career of astronomical science.

LECTURE VIII.

THE COMETARY WORLD.

THE wonderful characteristics which mark the flight of comets
through space ; the suddenness with which they blaze forth ;
their exceeding velocity, and their terrific appearance ; their
eccentric motions, sweeping towards the sun from all regions
and in all directions ; have rendered these bodies objects of
terror and dread in all ages of the world. While the planets
pursue an undeviating course round the sun, in its orbits nearly
circular, and almost coincident with the plane of the earth's

orbit, all revolving harmoniously in the same direction, the comets perform their revolutions in orbits of every possible eccentricity, confined to no particular plane, and moving indifferently in accordance with, or opposed to, the general motion of the planets. They come up from below the plane of the ecliptic, or plunge downwards towards the sun from above, sweep swiftly round this their great centre, and with incredible velocity wing their flight far into the fathomless regions of space, in some cases never again to re-appear to human vision.

In the early ages of the world, superstition regarded these wandering fiery worlds with awe, and looked upon them as omens of pestilence and war; and indeed, even in modern times, no eye can look upon the fiery train spread out for millions of miles athwart the sky, and watch the eccentric motions of these anomalous objects, without a feeling of dread. The movements of the planets inspire confidence. They are ever visible, and true to their appointed times, while the comet, erratic in its course, bursts suddenly and unannounced upon the sight, and no science can predict in the outset its uncertain track; whether it may plunge into the sun, or dash against one of the planetary systems, or even come into collision with our own earth, is equally uncertain, until after a sufficient number of observations shall have been made to render the computation of the elements of its orbit possible.

Previous to the discovery of the law of universal gravitation, comets were looked upon as anomalous bodies, of whose motions it was quite impossible to take any account. By some philosophers they were regarded as meteors kindled into a blaze in the earth's atmosphere, and when once extinguished they were lost for ever. Others looked upon them as permanent bodies, revolving in orbits far above the moon, and re-appearing at the end of long but certain intervals. When, however, it was discovered that, under the influence of gravitation, any revolving world might describe either of the four curves, the circle, ellipse, parabola, or hyperbola, it at once became manifest that the eccentric movements of the comets might be perfectly represented by giving to them orbits of the parabolic or hyperbolic form, the sun being located in the focus of the curve. According to this theory, the comet would become visible in its approach to its perihelion, or nearest distance from the sun,—would here blaze with uncommon splendour, and in its recess to the remote

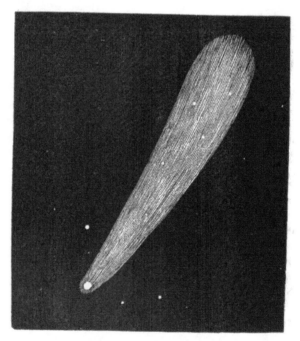

THE COMET OF 1680.   (Page 149.)

parts of its orbit, would gradually fade from the sight, relaxing its speed, and performing a large proportion of its vast curve far beyond the reach of human vision.

Such was the theory of Newton, and such were the fair deductions from the great law of nature which he had revealed to the world. He awaited with deep interest the appearance of some brilliant comet, whose career he might trace, in the full confidence that observation would confirm the truth of his bold hypothesis. Fortunately, his impatience was soon gratified. In the year 1680, a most wonderful comet made its appearance, which, by its splendour and swiftness, excited the deepest interest throughout the world. It came from the regions of space immediately above the ecliptic, and plunging downwards with amazing velocity, in a direction almost perpendicular to this plane, it appeared to direct its flight in such a manner that it must inevitably plunge directly into the sun. This was not, however, the case. Increasing its velocity as it approached the sun, it swept round this body with the speed of a 1,000,000 miles an hour, approaching the sun to within a distance of its surface of a sixth part of the sun's radius. It then commenced its recess, throwing off a train of light which extended to the enormous distance of 100,000,000 miles. With the swiftness of thought almost, it swept away from the sun, and was gradually lost in the distant regions of space whence it came, and has never since been seen. Such were the general characteristics of the body to whose rapid motions Newton attempted to apply the law of universal gravitation.

Its positions were marked with all the accuracy which the instruments then in use permitted, and it was found that a parabolic curve could be constructed, which would embrace all the places of the comet. The great eccentricity of its orbit, and its vast period, amounting to nearly six hundred years, gave to the comet great interest, but rendered it an unfit object for successful analytical research. The great English astronomer Halley, had studied it with the closest care, and with a rigid application of Newton's theory, he reached results quite as satisfactory as the circumstances of the case rendered possible.

Fortunately, in 1682, another comet made its appearance, to the study of which Halley devoted himself with a zeal and success, which has justly stamped his name on this remarkable body ; and as our limits forbid an extensive investigation of th history and theory of comets, I propose to examine this one with that degree of detail, which may convey some idea of the limits of human knowledge in this complicated department of science.

At the suggestion of Newton, Halley had searched all ancient and modern records, for the purpose of rescuing any historical details touching the appearance and aspect of comets, from the primitive ages down to his own time. On the appearance of the comet of 1682, he observed its position with great care, and with wonderful pains computed the elements of its orbits. He found it moving in a plane but little inclined to the ecliptic, and in an ellipse of very great elongation. In its aphelion, it receded from the sun to the enormous distance of 3,400,000,000 of miles. He discovered that the nature of its orbit was such as to warrant the belief that the comet would return at regular intervals of about seventy-five years ; and recurring to his historical table of comets, he found it possible to trace it back with certainty several hundred years, and with probability even to the time of the birth of Mithridates, one hundred and thirty years before Christ. At this, its first recorded appearance, its magnitude must have been far beyond anything subsequently seen, as its splendour is said to have surpassed that of the sun.

In the years 248, 324, and 399, of the Christian era, remarkable comets are recorded to have appeared, and the equality of interval corresponds well with Halley's comet. In the year 1006, it presented a frightful aspect, exhibiting an immense curved tail in the form of a scythe. In 1456, its appearance spread consternation through all Europe, and led to most extravagant acts on the part of the reigning pontiff, who actually instituted a form of prayer against the baleful influence of the comet, and thus increased the terrors of the ignorant and superstitious. The comet appeared with certainty in 1531, and again in 1607 ; and from an examination of all the facts, and with full confidence in his computations, Halley ventured the bold prediction, that this comet would re-appear about the close of 1758, or the beginning of 1759.

This was certainly the most extraordinary prediction ever made, and the distinguished philosopher, knowing that he could not live to witness the verification of this prophetic announcement, expresses the hope that when the comet shall return, true to his computed period, posterity will do him the justice to remember that this first prediction was made by an Englishman. In the age when these investigations were made, the theory of comets was in its infancy; and it is believed by those competent to form a just opinion, that Halley was the only man living who could have computed the orbit of his comet.

As the period approached for the verification of this extraordinary announcement, the greatest interest was manifested among astronomers, and efforts were made to predict its coming with

Dr. Halley.

greater accuracy, by computing the disturbing effects of the larger planets within the sphere of whose influence the comet might pass. This was a new and difficult branch of astronomical science, and it would be impossible to convey the least idea of the enormous labour which was gone through by Clairaut and Lalande, in computing the perturbations of this comet through a period of two revolutions, or one hundred and fifty years.

"During six months," says Lalande, "we calculated from morning till night, sometimes even at meals; the consequence of which was, that I contracted an illness which changed my constitution during the remainder of my life. The assistance rendered by Madame Lepaute was such, that without her we never should have dared to undertake the enormous labour, in which it was necessary to calculate the distance of each of the two planets, Jupiter and Saturn, from the comet, separately for every degree, for one hundred and fifty years."

Amid all these difficulties, the computers toiled on, until finally, the period coming on rapidly for the comet's return, they were forced to neglect some minor irregularities, and Clairaut announced that the comet would be retarded one hundred days by the influence of Saturn, and five hundred and eighteen days by the action of Jupiter ; he therefore fixed its perihelion passage for the 13th of April, 1759, stating, at the same time, that the result might be inaccurate by some thirty days either way, in consequence of being pressed for time, and his having neglected certain small perturbations.

These results were presented to the Academy of Sciences on the 14th of November, 1758, and on the 25th of the following December, George Palitch, an amateur peasant astronomer, caught the first glimpse of the long expected wanderer, which, after an absence of three-quarters of a century, once more returned to crown with triumph the great English astronomer who first foretold its period, and the eminent French mathematicians, who had actually computed its perihelion passage to within nineteen days in seventy-six years !

Here, then, was a new world added to the solar system, linked to the sun by the immutable law of gravitation : sweeping out into space to the amazing distance of 3,800,000,000 of miles ; lost to the gaze of the most powerful telescope, and yet traced by the human mind through its vast and hidden career, with an accuracy and precision, from which there was no escape. The very small error of nineteen days in Clairaut's computation strikes us with astonishment, when we remember the imperfect state of analytical science at that day, and the fact that two planets, Uranus and Neptune, which have since been discovered, were then not even suspected to have any existence.

The magnificent display which had marked some of the early returns of Halley's comet, and which produced such consternation among all classes, the educated as well as the ignorant, were not presented during its appearance in 1758. This was owing in part, to the unfavourable position of the earth in its orbit at the time of the comet's perihelion passage. The vast trains of light which are sometimes seen to accompany comets are only displayed in their approach to the sun. They attain their

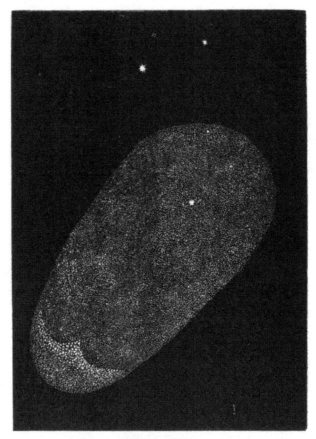

HALLEY'S COMET.

As seen by Sir John F. Herschell, October 29th 1835. (Page 153.)

Strange, and almost incredible as it must appear, guided by these predictions, M. Dumouchel, director of the observatory of the Roman College, on the evening of the 5th of August, 1835, fixed his telescope in the position indicated by computation, and, on placing his eye to the tube, lo ! the comet appeared, as a faint and almost invisible stain of light on the deep blue of the heavens.

Again did science triumph, in the most remarkable manner, and the computed orbit of the comet was followed by it with the most surprising accuracy. The perihelion passage was predicted to within nine days of its actual occurrence ; a most astonishing approximation to the truth, when it is remembered that this body, far as it penetrates into space, never, even at the remotest point of its orbit, escapes from the sensible influence of the planet Jupiter. Moreover, at that time, the new planet, Neptune, was unknown, and its influence over the comet could not be taken into account.

It is interesting to remark the confidence with which astronomers relied on Halley's comet for information relating to those bodies which inhabited the regions of space exterior to the known limits of the solar system. It was urged by every computer that the orbit of this comet would one day come to be so perfectly known, that the perturbations due to the recognised bodies of the solar system might be computed with such precision, that the residual perturbations might be pronounced to be the effect of unknown planets or comets circulating in the distant regions of space. This conjecture has been realised, although by different means, and a planet is now added to our system, which revolves in an orbit so vast as to circumscribe within its limits the entire sweep of the comet ; and as the orbits of Neptune and Halley's comet are inclined under an angle of only 15° or 16°, a time will come when the perturbations experienced by the comet when at its aphelion, from Neptune's influence, may be so great, that, but for the fortunate discovery of the cause, would have falsified, in the most unaccountable manner, the predictions of the comet's return by future astronomers. During its late return, the finest telescopes in the world were employed in a critical examination of the physical condition of Halley's comet. Elaborate drawings have been made by M. Struve, the distinguished director of the imperial observatory at Pulkova, Russia, with the grand refract-

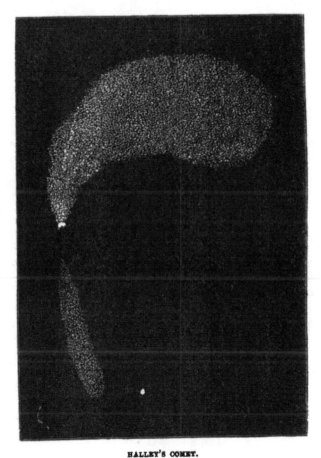

HALLEY'S COMET.

As seen by Struve, October 12th, 1835. (Page 155.)

ing telescope under his charge, and also by Sir John Herschel, at the Cape of Good Hope, with a 20-feet reflector of superior power. To these beautiful drawings reference will be made hereafter.

The most wonderful changes in the magnitude and figure of the comet were observed to take place from night to night, and almost percepti- ble from hour to hour, under the eye of the ob- server. The nucleus was sometimes seen sharp and strongly condensed, with more or less nebulous light around it. Sometimes a luminous crescent became distinctly visible near the nucleus, giving to the comet a most extraordinary ap- pearance. At one time M. Struve saw the comet at- tended by two delicately- shaped appendages of light of a most graceful form, the one preceding, the other following the nucleus of

Sir J. F. Herschel, Bart., F.R.S.

the comet. At other times it was seen to be surrounded by a sort of semicircular veil, which, extending backwards, was lost in a double train of light, which flung itself out to a vast distance from the body of the comet.

Leaving, for the present, the consideration of the physical constitution of these eccentric bodies, we proceed to the exami- nation of a remarkable object, which bears the name of Encke's comet, in consequence of the discovery by this learned astro- nomer that its orbit was elliptical, and its period of revolution so short, as to fall fairly within the limits of perpetual examination.

In 1818, a comet was discovered by Pons, not at all remark- able for its magnitude, for it was even invisible to the naked eye ; but when the attempt was made to represent its places by a parabolic orbit, which had thus far been invariably applied to

the comets, it was found impossible to assign any elongated
orbit which would embrace the observed positions of the comet.
After a very elaborate investigation, Professor Encke at length
reached the conclusion that the orbit was not a parabola, but an
ellipse of comparatively small dimensions, and that this comet
was actually revolving around the sun in a period of about three
years. This discovery excited a great deal of interest, for it
was the first in which a short period had been detected, and
efforts were at once made to identify the new member of the
solar system in its preceding revolutions. Olbers determined
its identity with a comet which appeared in 1795, and subsequently ascertained that another, which had been observed but
twice in 1786, and from which observations no elements could
be computed, could be no other than the new comet of that
period. In this way, observations on this interesting object
were obtained, stretching through some thirty-three years, or
about ten of its revolutions. This extended series of observations furnished the data for a critical examination of the elements
of the comet's orbit, and Professor Encke, having discussed them
with elaborate care, reached the astonishing conclusion, that the
magnitude of the orbit was gradually diminishing, the periodic
time growing shorter from revolution to revolution, and that
the comet was certainly falling nearer and nearer to the sun !

To account for this extraordinary phenomenon, the learned
astronomer, having exhausted all causes known to exist in the
solar system, finally, with much hesitation, announced the theory
of the comet's motion in a *resisting medium*. The existence of
such a medium was in direct opposition to all the received
doctrines of astronomy, and the absolute necessity for its use in
this instance was looked upon by astronomers with feelings of
strong distrust. But Encke argued that such a medium might
exist, of such exceeding tenuity, as not sensibly to affect the
movements of the ponderous planets, while a filmy mass of
vapour, such as this comet undoubtedly was, might be very
sensibly retarded in its original velocity, which would diminish
continually the centrifugal force, and give to the central attraction of the sun a constantly increasing power, which would
produce precisely the phenomena exhibited by the comet.

With these views, Encke predicted the re-appearance of the
comet in 1822. In consequence of its great southern declination
at that period, it escaped all the European observers, and was

THE GREAT COMET OF 1744. (Page 157.)

only seen at Paramatta, by Rumker. The approach to the sun was in some degree confirmed by these observations, but it was impossible to reconcile all the observations with the hypothesis of a medium of given density. The return in 1825 was not favourable for deciding the question, which had now become one of the deepest interest.

Its re-appearance in 1828-9 was awaited with great anxiety by the friends and opponents of the new theory. The comet came round, and passed its perihelion approximately in accordance with the predictions, but the discrepancies from 1819 up to 1829, with any theory, were so great, as to give much perplexity to those engaged in the computations. After long and patient examination, the cause of this difficulty was finally detected. The plane of the comet's orbit makes but a small angle with the orbit of Jupiter, and when the comet is in aphelion, or farthest from the sun, it always approaches very near to the path described by the planet.

A time may then come when Jupiter shall be in the act of passing that part of its orbit very near to the aphelion point of the cometary eclipse, while the comet occupies its aphelion, bringing these bodies into close proximity, and producing excessive perturbations in the movements of this almost spiritual mass. Such, indeed, was the configuration between the returns of 1819 and 1829, on which occasion the comet was delayed in its return to its aphelion by nearly nine days, by the powerful attraction of Jupiter. Under these circumstances, any error in the assumed mass of the planet would exhibit itself in an exaggerated form in the perturbations of the comet. But it was believed in the outset of this investigation, that the mass of Jupiter, employed by Laplace in his theory of the planets, and computed by Bouvard, could be relied on as accurate. Indeed, Laplace had applied the calculus of probabilities, and had found that there was but one chance out of eleven millions that the mass he had adopted could be in error by the one hundredth part of its value.

Suspicion, however, having been aroused with reference to the mass of Jupiter, efforts were at once commenced to sift thoroughly the matter, and three different computers of high reputation undertook the determination of Jupiter's mass by different processes. Encke obtained a mass from the perturbations of the small planet Vesta, Nicolai from the perturbations of Juno, and

Airy re-examined the original measures of the elongations of Jupiter's satellite, made new measures, and thus obtained new data for the resolution of the problem of Jupiter's mass. The results obtained by the three astronomers agreed in a most remarkable manner, and proved incontestably that Laplace's value of the mass of this planet was in error more than *four* times the hundredth part of its value, and that, instead of requiring 1,070 globes of the magnitude of Jupiter to balance the sun, only 1,049 were necessary.

With the new mass of Jupiter it seemed possible, by admitting a resisting medium, to account for all the perturbations of Encke's comet, and for a time this theory seemed to receive greater consideration from distinguished men. The appearance of Halley's comet in 1835 again threw great doubt over the subject, for it was found impossible to reconcile the movements of the two comets with any assumed density of a resisting medium. Some have been disposed to adopt the idea that the revolution of the planets for ages in the same direction, in this supposed ethereal fluid, has impressed upon it a certain amount of motion in the same direction, and that those comets which chance to revolve with the current will be found to be operated upon differently from those which may happen to come into our system in a direction opposed to the current.

I confess, frankly, that my own mind has always revolted against the doctrine of a resisting fluid. There are so many ways in which the single phenomenon of the gradual approach of Encke's comet to the sun may be accounted for, without resorting to an hypothesis which involves the entire destruction of the planetary system, whose perpetuity has been so effectually provided for by the great Architect of the universe, that it would require the most unequivocal testimony to secure the full consent of my own mind to the adoption of this remarkable theory. It is proper, however, to say, that it has long been received with favour by men to whose judgment I am generally disposed to yield with implicit confidence.

Leaving the further consideration of this subject for the present, we proceed to the examination of another comet of short period, which has excited great attention, especially in its recent return. As early as 1805, Professor Gauss, in computing the elements of the orbits of the comets of that year, found one which seemed to complete its revolution in about six years.

This comet, however, was lost sight of, and it was not until 1826 that M. Biela discovered the same comet on its return to its perihelion. This discovery appears to have been the result of computation, but how far the investigation was carried I have never been able to learn.

The same object was also discovered by M. Gambart about the same time, who, on fixing its elements, found that it performed its revolution about the sun in an ellipse, with a period of six and three-quarter years. This comet, like Encke's, is only to be seen with the telescope. It presents no solid, or even well defined nucleus, and appears to be a mere vapoury mass, of exceeding tenuity. Taking into account the disturbing influence of Jupiter, the returns of Biela's comet, as predicted, agreed well with observation, and gave confidence in the theory on which the predictions were founded.

The return in 1832 excited the liveliest interest throughout the civilised world, in consequence of the fact that it was discovered from computation, that on the night of the 29th of October, this comet would pass a little within the earth's orbit, and those unacquainted with the subject received the impression from this announcement, that the earth and comet would come into collision, producing the most terrific consequences. Such was the consternation excited, throughout the city of Paris especially, that the Academy of Sciences found it necessary to give to the subject their serious attention, and finally gave the matter in charge to M. Arago, who produced an elaborate report on the subject of comets gener-

M. Arago.

ally, which served to calm the popular apprehension, and has proved to be a valuable addition to our knowledge on this difficult subject.

In this report, M. Arago showed that the comet would indeed cross the earth's track at the time predicted, but at the moment of crossing, the earth would be some fifty-five millions of miles distant from the point occupied by the comet, and could not experience the slightest possible influence from such a body, at such a distance.

If the comet had been delayed in its approach for thirty days by any disturbing cause, then indeed the earth and comet would have filled at the same time the point where their orbits intersect, and the dreaded collision would have taken place. The consequences of such a shock it is impossible to conjecture; but reasoning from the known physical condition of the comet, none of the terrible disasters so generally anticipated would have occurred. The exceeding rarity of the matter composing this body may be inferred from the statements of Sir John Herschel. " It passed," says he, " over a small cluster of most minute stars of the 16th and 17th magnitude; and when on the cluster presented the appearance of a nebula resolvable, and partly resolved; the stars of the cluster being visible through the comet. A more striking proof could not have been offered of the extreme translucency of the matter of which the comet consists. The most trifling fog would have effaced this group of stars, yet they continued visible through a thickness of cometic matter, which, calculating on its distance and apparent diameter, must have exceeded 50,000 miles, at least toward its central parts. That any star of the cluster was *centrally* covered, is indeed more than I can assert; but the general bulk of the comet might be said to have passed centrally *over the group*."

Such is the nature of the body from whose contact the ignorant apprehended the most fearful convulsions. Olbers, who studied the subject with great care, was disposed to think that in case the earth had passed directly through the comet, no inconvenience would have occurred, and no change beyond a slight influence on the climate would have been experienced.

It is useless to speculate with reference to the probable consequences of a collision, which there is scarcely one chance in millions can ever occur. Science has as yet discovered no guarantee for any planet against the possible shock of a comet; but an examination of the delicate adjustments of our own system, and those of Jupiter and Saturn, would seem to indicate to us that in all past time no derangement has ever occurred from such a cause.

The last return of Biela's comet was marked by a phenomenon unexampled, so far as I know, in the history of these wandering bodies. True to the predictions of Santini, the comet first became visible on the evening of the 26th of November, 1845, and in the precise point which had been assigned by theory. De Vico, the director of the observatory at Rome, was the first to catch a glimpse of the expected comet. Nothing remarkable in its appearance was noticed until about the 29th of December, when Mr. E. C. Herrick, of New Haven, pointed out to several friends what he regarded to be a sort of anomalous tail, but shooting out from the head of the comet in a direction entirely at variance with the usually received theory, that the tail is always opposite to the sun. In this supposed tail a kind of *knot* was noticed, brighter and more condensed than any other part. Owing to insufficient optical power, the true character of the phenomenon was not fairly detected by Mr. Herrick.

On the night of the 12th of January, 1846, Lieut. Maury, in charge of the observatory at Washington, United States, discovered that what had hitherto appeared as a single body, was actually composed of *two distinct and separate comets*. This most extraordinary fact was immediately announced, and the double character was observed at all the principal observatories in Europe and the United States. There can be no doubt whatever as to the reality of the appearance. The comet actually became double, and the two parts, bound together by some inscrutable bond, continued their swift journey through space, pursuing almost exactly the route predicted for the single comet.

From measures obtained by Professor Challis, of Cambridge, England, on the 23rd of January, 1846, the two comets were separated from each other by a distance equal to about one-thirteenth the apparent diameter of the sun. On the 28th of the same month, Sir John Herschel records the following notices :—" The comet was evidently double, consisting of two distinct nebulæ, a larger and a smaller one, both round, or nearly so ; the one in advance faint and small, and not much brighter in the middle ; the one which followed nearly three times as bright, and one and a half times larger in diameter, and a good deal brighter in the middle, with an approach to a stellar point."

On the evening of the 9th of February, having returned to the observatory at Cincinnati, after an absence of more than two

M

months, I had an opportunity of beholding, for the first time, these wonderful objects, with the twelve-inch refractor. The moon was nearly full, and yet the comets were distinctly visible, both included within the limits of the field view of the instrument, and separated from each other by a distance equal to about the eighth part of the sun's diameter. The preceding comet was evidently the brighter of the two.

Clouds prevented a continuous examination of the comets from night to night, but on the evening of the 21st of February, I was surprised to find a remarkable change in the relative brilliancy of the two parts. On that evening the following comet was very decidedly brighter than its companion, and from observations made elsewhere, the change of relative brightness seems to have been effected about the 13th or 14th of February. The change was observed by Professor Encke, of Berlin, as early as the 14th. On the evening of the 21st of February, both comets exhibited distinct trains of light, extending from the sun, and in directions parallel to each other. The centre of the nucleus of each comet was brighter than the surrounding portions, but there was no stellar point visible. The nebulosity of the two points did not intermingle.

The distance between the comets increased from day to day, until, on the 25th of February, they were separated by an amount equal to 445 seconds of arc, or between a fourth and fifth part of the sun's diameter. A part of the increase of distance was only apparent, arising from the approach of the comets to the earth, but the comets were actually receding from each other while pursuing their rapid flight through space.

Neither did the line joining the central points of the comets remain parallel to itself. From the 23rd of January to the 11th of February, this line shifted its position by an amount of angular motion equal to 8°, as is shown by a comparison of the measures of Challis and Encke. By the 21st of February, this angular motion had been nearly destroyed by a retrograde movement, and thus the comets were seen to oscillate about each other according to some mysterious law which has never been revealed. Such is a brief sketch of the phenomena presented by Biela's comet in its late return. Its next appearance will be looked for with deep interest, to confirm or destroy certain theories which have been propounded to explain its duplex character.

While the periods of the comets which we have thus far con-

THE COMET OF 1811. (Page 163.)

sidered are comparatively short, those of others which have visited our system have been ascertained to extend to many thousands of years. The great comet of 1811, one of the most brilliant of modern times, in consequence of its remaining visible for nearly ten months, gave ample opportunity for the investigation of the elements of its orbit. After a careful investigation, M. Argelander fixes its period of revolution at 2,888 years Bessel had examined the same subject previously, and probably with less attention, but obtained a period even greater than Argelander's, amounting to 3,383 years.

The comet of 1807 also occupied the attention of Bessel. A long series of observations furnished the data for computing its elements. The periodic time was fixed at 1,543 years. These computations are necessarily only approximate. The difficulty of obtaining accurately the periodic time increases with the length of the period, and all that can be done is to fix a limit below which it cannot fall. These vast periods give to us the means of learning somewhat of the great distance to which these objects penetrate into space. The comet of 1811, having a period probably three thousand times greater than that of our earth, must revolve at a mean distance from the sun of more than 80,000,000,000 miles, and in consequence of its very near approach to the sun at its perihelion, its greatest distance cannot fall below 160,000,000,000 miles !

Great as this distance is, it is perfectly certain that there are many comets which revolve in orbits far more extensive than the one described by the comet of 1811. Indeed, there seems to be no limit to the distance to which these bodies may sweep outward from the sun ; and their return depends simply on the fact whether they recede so far as to fall within the attractive influence of some other sun, towards which they begin to urge their flight, and through whose system of planets they carry the same apprehensions of danger which have been caused in our own.

In reflecting on these singular objects, we are led to inquire what they are, whence their origin, and by what laws are the vast trains of light which occasionally distinguish them developed. Arago divides comets into three classes, with reference to their physical constitution. He thinks they occasionally appear round, and with well defined planetary discs, showing them to be solid opaque bodies, in all respects resembling planets, and only differing from these in the great eccentricity of their

M 2

orbits. In confirmation of this opinion, he asserts that comets have been seen to transit the sun, and when passing between this luminary and the eye of the spectator, they appear round and black, like the planets Mercury and Venus, when seen under the same circumstances. An example of this kind occurred on the 18th of November, 1826, when the transit of a comet across the sun was witnessed by two persons, widely separated from each other.

A second class of comets comprehends those in which there is a nucleus, but devoid of opacity, permitting the light to penetrate through even that portion which may possibly be solid. The third class, and that by far the most numerous, comprehends those comets destitute entirely of any solid nucleus, consisting of matter so attenuated as to compare fairly with nothing of which we have any knowledge on the earth's surface. The comets named from Encke and Biela appear to belong to this class ; and even Halley's comet, according to the opinion of Sir John Herschel, seems, at its last return, to have been entirely turned into vapour in its perihelion passage.

No theory, with the exception of Laplace's nebular hypothesis, has ever been framed to explain the origin of these wandering bodies. This is not the place to enter into a full development of this subject. A few hints only can be given. Laplace, following up the speculation of Sir William Herschel, applied the theory of that astronomer to the formation of the solar system, comprehending the comets, as well as the planets and their satellites. This theory supposes that the original chaotic condition of the matter of all suns and worlds was nebulous, like the matter composing the tails of comets. Under the laws of gravitation, this nebulous fluid, scattered throughout all space, commences to condense towards certain centres. The particles moving towards these central points, not meeting with equal velocities, and in opposite directions, a motion of rotation is generated in the entire fluid mass, which in figure approximates the spherical form.

The spherical figure once formed, and rotation commenced, it is not difficult to conceive how a system of planets might be produced from this rotating mass, corresponding, in nearly all respects, to the characteristics which distinguish the planets belonging to our own system. If, by radiation of heat, this nebulous mass should gradually contract in size, then a well-known law o rotating bodies would insure an increased velocity of rotation. This might continue until the centrifugal force, which increases rapidly with the velocity of the revolving body, would finally come to be superior to the force of gravity at the equator, and from this region a belt of nebulous fluid would thus be detached, in the form of a ring, which would be left in space by the shrinking away of the central globe. The ring thus left would generally coalesce into a globular form, and thus would present a planet with an orbit nearly, if not quite, circular, lying in a plane nearly coincident with the plane of the equator of the central body, and revolving in its orbit in the same direction in which the central globe rotates on its axis.

As the globe gradually contracts, its velocity of rotation continually increasing, another ring of matter may be thrown off, and another planet formed, and so on, until the cohesion of the particles of the central mass may finally be able to resist any further change, and the process ceases.

The planetary masses, while in the act of cooling and condensing, may produce satellites in the same manner, and by the operation of the same laws by which they were themselves formed. Strange and fanciful as this speculation may appear, there are many facts which tend strongly to give it more than probability. It accounts for all the great features of the solar system, which, in its organisation, presents the most indubitable evidence that it has resulted from the operation of some great law.

The sun rotates on an axis in the same direction in which the planets revolve in their orbits; the planets all rotate on their axes in the same direction; they all circulate around the sun, in orbits nearly circular, in the same direction, and in planes nearly coincident with the plane of the sun's equator. The satellites of all the planets, with one single exception, revolve in orbits nearly circular, but little inclined to the equators of their primaries, and in the same direction as the planets. So far as their rotation on axes has been ascertained, they follow the general law. In one instance alone we find the rings of matter have solidified in cooling, without breaking up or becoming globular bodies. This is found in the rings of Saturn, which present the very characteristics which would flow from their formation according to the preceding theory. They are flat and thin, and revolve on an axis nearly, if not exactly, coincident with the axis of their planet. Their stability, as we have seen, is guaranteed by conditions of wonderful complexity and delicacy, and the adjustment of the rings to the planet, (humanly speaking,) would seem to be impossible after the formation of the planet. At least it is beyond our power to conceive how this could be accomplished by any law of which we have any knowledge, and we must refer their structure at once to the fiat of Omnipotence.

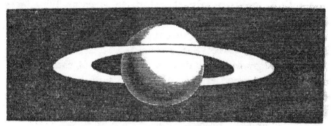

Telescopic appearance of Saturn in December, 1850.
(Scale, 20 seconds of arc to one inch.)

Granting the formation of a single sun by the nebular theory, and we account at once for the formation of all other suns and systems throughout all space ; and according to the advocates of this theory, the comets have their origin in masses of nebulous matter occupying positions intermediate between two or more great centres, and held nearly in equilibrio, until, finally, the attraction of some one centre predominates, and this uncondensed filmy mass commences slowly to descend towards its controlling orb. This theory would seem to be sustained, so far as a single truth can sustain any theory, by the fact that the comets come into our system from all possible directions, and pursue their courses around the sun either in accordance with or opposed to the directions in which the planets circulate. Their uncondensed or nebulous condition results from the feeble central attraction which must necessarily exist in bodies composed of such small quantities of matter. Moreover, in some cases at least, there is reason to believe, that in their perihelion passage they are entirely dissipated into vapour by the power of the sun's heat, and may thus revolve for ages, going through alternate changes of solidification and evaporation.

But whence come the enormous trains of light which sometimes attend these wandering bodies ? The last return of Halley's comet has furnished the data for the positive illustration of this mysterious subject. Sir John Herschel, after a careful and most elaborate examination of all the physical characteristics of this comet, comes to the conclusion that the figure of the comet, envelope and tail, could not be a figure of equilibrium under the law of gravitation. He is therefore compelled to bring in a *repulsive force* to explain the phenomena.

I cannot do better than to quote his own language in this bold introduction of a new power. " Nor let any one be startled at the assumption of such a repulsive force as here supposed. Let it be borne in mind that we are dealing, in the tails of comets, with phenomena utterly incompatible with our ordinary notions of gravitating matter. If they be material in that ordinarily received sense which assigns to them only inertia and attractive gravitation, where, I would ask, is the force which can carry them round in the perihelion passage of the nucleus, in a direction pointing continually *from* the sun, in the manner of a rigid rod, swept round by some strong directive power, and in contravention to all the laws of planetary motion, which would require a slower angular motion of the more remote particles, such as no attraction to the nucleus could give them, be it ever so intense ? The tail of the comet of 1680, in five days after its perihelion passage, extended far beyond the earth's orbit, having, in that brief interval, shifted its angular position nearly 150°. Where can we find, in its gravitation either to the sun, or to the nucleus, any cause for this extravagant sweep ?

" But, again, where are we to look, if only gravity be admitted, for any reasonable account of its projection *outward from the sun*, putting its angular motion out of the question ? Newton calculated that the matter composing its upper extremity quitted the nucleus only two days previous to its arriving at this enormous distance."

Herschel argues the inadequacy of gravitation to account for these wonderful phenomena. The velocity with which the matter composing the tail shot forth from the head of the comet, *from the sun*, was far greater than that which the sun could impress on a body falling to it, even from an infinite distance. An energy of a different kind from gravitation, and far more powerful, must exist, to produce such results. If, then, we are forced to the admission that a power exists in the sun capable of repelling matter of a certain quality existing in comets, a way is opened for the explanation of some of the most difficult problems with which the mind has been obliged to contend.

The diminution of the periodic time of Encke's comet has led some astronomers to adopt the idea of the existence of a resisting medium. But in case the sun possesses the power of repelling the matter of comets in their perihelion passage, a part of the matter thus repelled may be driven entirely beyond the attractive influence of the nucleus, and be irrecoverably lost. In this case, a diminution of mass would inevitably involve a like diminution of periodic time, a contraction of the orbit, and all the phenomena presented by this mysterious object. Herschel even thinks it possible, on this theory, to account for the separation of Biela's comet into two distinct objects, and it appears to me that it presents the most reasonable explanation of the luminous appearance seen at certain seasons of the year, called the *zodiacal light*. This phenomenon appears to be a ring of nebulous matter surrounding the sun, and some of whose particles are sustained at a much greater distance than could be accounted for by gravitation. Admitting the repulsive power already adverted to, there is no difficulty in understanding how this nebulous ring may be sustained at a vast distance from the sun.

Here we freely admit that we enter the confines of the unknown. We have left the solid ground of truth and certainty, and are pushing our investigations into the dim twilight of the invisible and uncertain. But as antiquity predicted that the time would come when the comets would be traced in their career, their periods revealed, and their orbits ascertained, so we may confidently hope that, at no very distant day, all the mysteries which hang around these chaotic worlds will be fully revealed, and a knowledge of their physical condition shall reward the long study and deep research of the human mind.

LECTURE IX.

HUS far our attention has been directed to an examination of the achievements of the human mind within the limits of our own peculiar system. We have swept outward from the sun through the planetary worlds, until we have reached the frontier limits of this mighty family. Standing upon the latest found of all the planets, at a distance of more than 3,000,000,000 of miles from the sun, we are able to look backwards, and examine the worlds and systems which are all embraced within the vast circumference of Neptune's orbit.

An occasional comet, overleaping this mighty boundary, and flying swiftly past us, buries itself in the great abyss of space, to return after its "long journey of a thousand years," and report to the inhabitants of earth the influences which have swayed its movements in the invisible regions whither it speeds it flight.

The magnificence and complexity of the great system of planets, and satellites, and comets, which constitute the sun's retinue, the immense magnitude of some of these globes, their periods of revolution, and reciprocal action, would seem to furnish a sufficient exercise, not only for the highest intellectual efforts, but for the entire energy which the human mind can exert. But the whole of this stupendous scheme, as we shall soon see, is but an infinitesimal portion of the universe of God, one unit among the unnumbered millions which fill the crowded regions of space. Standing at the verge of the planetary system, we find ourselves surrounded by a multitude of shining orbs, some radiant with splendour, others faintly gleaming with beauty. The smallest telescopic aid suffices to increase their number in an incredible degree, while with the full power of the grand instruments now in use, the scenes presented in the starry heavens become actually so magnificent as to stun the imagination and overwhelm the reason. Worlds and systems, and schemes and clusters, and universes, rise in sublime perspective, fading away in the unfathomable regions of space, until even thought itself fails in its efforts to plunge across the gulf by which we are separated from these wonderful objects.

In our measurements within the limits of the solar system, the radius of the earth's orbit has sufficed for a unit with which to exhibit the distances of the planets and comets. Great as is this unit, measuring no less than 95,000,000 miles, we shall soon find it far too minute and insignificant to serve in our researches with reference to the grand scale of the visible universe. To obtain comprehensible ideas with reference to the interstellar spaces, we shall be obliged to call to our aid a unit, not exactly of distance, but of velocity; and before entering on the full exhibition of the main object of this lecture, permit me to direct your attention to a remarkable discovery, by which the important fact has been revealed, that light does not pass instantly from a luminous body to any remote object on which it may fall, but with a progressive motion, whose actual velocity has been ascertained. The important bearing of this discovery will become apparent as we advance in our examination of the sidereal heavens.

After the motions of the four moons of Jupiter had been sufficiently observed to construct tables of their movements, with a view to predict their eclipses, some unaccountable phenomena presented themselves, which, for a long time, baffled all efforts to explain them. It should be remembered, that the orbit of Jupiter encloses that of the earth, and when the two planets happen to be on the same side of the sun, and in a straight line passing through that orb, they are then at their least distance from each other, and are said to be in conjunction. Now suppose Jupiter to remain stationary, at the end of half a year the earth will

l
s
o
r
v
s
t
n
g
h
t
s
f
t
v

r
e
u
i
v
c
s
o
e
y
f
l
v
t
ε
l

(
v
]
(
(
(
t
(
t

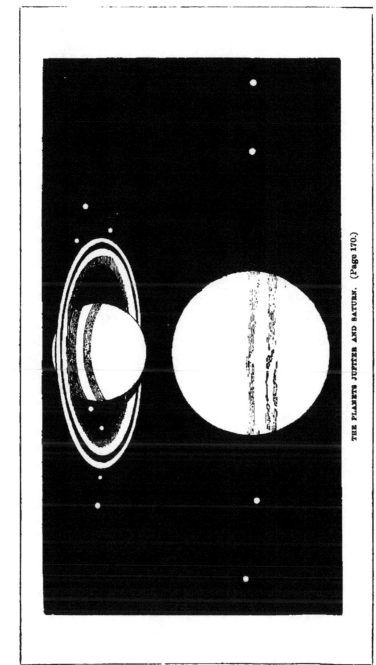

THE PLANETS JUPITER AND SATURN. (Page 170.)

have reached the opposite point of her orbit, and will now be more distant from Jupiter by an amount equal to the diameter of her orbit, or nearly 200,000,000 miles. Retaining carefully these positions in the mind, we shall follow the facts about to be presented with the greatest ease.

It was found that these eclipses of Jupiter's satellites, which occurred while the earth and planet were at their least distance from each other, always came on *sooner* than the time predicted by the tables; while, on the contrary, those which took place when the planets were most remote from each other, occurred *later* than the computed time. A still more extended examination of these remarkable phenomena demonstrated the fact, that the discrepancies depended evidently on the absolute increase and decrease of distance which marked the relative position of the planets in their revolutions around the sun. For a long time, no explanation of these undeniable truths could be found, until the mystery was finally solved by Roemer, a Danish astronomer, who, with admirable sagacity, traced these irregularities to their true source, and found that they arose from the fact that light travelled through space with a finite and measurable velocity.

The explanation is simple. When Jupiter and the earth are at their least distance from each other, the stream of light flowing from the satellite of the great planet traverses a shorter space to reach the eye of the observer on the earth, by nearly 200,000,000 miles, than when the planets are most remote from each other. In case, therefore, this stream is in any way cut off, it will run out sooner in the first than in the second position, by the time required to pass over the diameter of the earth's orbit. The stream of light is actually shorter, by 200,000,000 miles, in the first than in the second position of the planets.

Now the satellites of Jupiter receive their light from the sun; they reflect this light to the earth, and when the body of their primary is interposed between them and their source of light, they are eclipsed; their light is cut off; its flow is interrupted; and when the stream of light starting from them at the instant the supply is cut off shall have run out, then, and not till then, does the satellite become invisible. This explanation accounted for all the phenomena in the most beautiful manner.

The tables had been constructed from the mean of a great number of observed eclipses. Hence, those which took place while Jupiter and the earth were near to each other, would happen earlier than prediction ; while those taking place when the planets were at their greatest distance, would occur later than the time given by the tables. But the velocity with which this mysterious, subtle, intangible substance called light, flew through the regions of space, as determined by this wonderful theory, was so great as to startle the minds of even its strongest advocates, and to demand the most positive testimony to induce the belief of those disposed to scepticism. It was found to traverse a distance equal to the entire diameter of the earth's orbit, or 190,000,000 miles, in about 16 minutes ! giving a velocity of 12,000,000 miles per minute, or 192,000 miles in each second of time !

It is not our purpose to enter into any investigation as to the true theory of light, whether it be an actual emanation from a luminous body of material particles, or whether it be a mere vibratory or undulating motion produced by luminous bodies on some ethereal medium. My only object, at this time, is to assert the undoubted fact, that in case a luminous body were to be suddenly called into being, and located in space at the distance of 12,000,000 miles from the eye of an observer, who was on the look out for its light, this light would not reach him until *one minute* after the creation of the object ; and should it suddenly be struck from existence, the same observer would behold it for one minute after the extinction.

Should any mind revolt from these statements, should the difficulty of the investigation, and the incredible velocity of light, demand higher and better evidence, before full faith can be given to the theory, I can only say that this evidence shall be given before we close this discussion, and with a fullness and clearness which shall set all doubts at defiance.

I now proceed to an examination of the great problem of the parallax of the fixed stars, a problem which has taxed the ingenuity of the greatest minds, and which has called into requisition the most admirable skill for a period of more than three hundred years. A familiar explanation of the nature of this problem may prepare the way for a rapid sketch of the various means which have been employed in its solution. If it were possible to measure on the earth's surface a base line of a thousand miles in length, by locating an observer at each extremity of this base, with instruments suitable to fix the moon's place among the fixed stars, the telescopes of those two observers, directed to the moon's centre at the same instant, would incline towards each other, and the visual ray from each of those instruments would meet at the moon's centre, and form an angle with each other. This angle or opening of the visual rays, is called the *parallax*, and in case the object under examination were a fixed star, then would the angle in question be called *the parallax of the fixed star*.

It is readily seen that when the length of the base is known, and the parallactic angle measured, then the length of the visual ray may be at once determined, and the distance of the object is made known by the simplest rules of geometry. Parallax, then, in general, is the *apparent* change in the place of an object, occasioned by the *real* change in the place of the spectator.

The whirling of the trees of a forest, produced by the rapid speed of the beholder along a railway, is a parallactic motion, and becomes less and less perceptible as the velocity of the spectator diminishes, or as the distance of the seemingly moving object becomes greater. To measure the distance of the fixed stars is then equivalent to determining the amount of the parallactic change in their relative positions, occasioned by the actual change of the positions from which they may be viewed by a spectator on the earth's surface.

With the sun and moon and planets, a base line equal to the earth's diameter, or about 8,000 miles, has sufficed to produce a sensible and measurable parallax; but when we extend our visual rays to a fixed star, from the extremities of this base, their directions, to our senses, are absolutely parallel, or, in other language, the parallax arising from such a base is perfectly insensible. This first effort indicates, at once, the vast distance of the objects under examination; for such is the accuracy with which minute spaces are now divided, that parallax may be detected in case the object is even 160,000 times farther distant than the length of the base line.

When the orbital motion of the earth was first propounded by Copernicus, and it was asserted to revolve in an ellipse of nearly 600,000,000 miles in circumference, and with a motion so swift that it passed over no less than 68,000 miles in every hour of time, the opponents of these startling doctrines exclaimed, No! this is impossible; for if we are sweeping around the sun in this vast orbit, and with this amazing velocity, then ought the fixed stars to whirl round each other, as do the forest trees to the traveller flying swiftly by them.

But the stars of heaven do not move. Seen from any point, and at any time, their places are ever the same, fixed, immutable, eternal; the bright and living witnesses of the extravagance and absurdity of this new and impossible theory! To this reasoning, which was well founded, and without sophistry, the Copernicans could only reply, that such was the enormous distance of the sphere of the fixed stars, that no perceptible change was occasioned by the revolution of the earth in its orbit. But this was mere assertion, and the opponents met the statement by this very plain exhibition of the case. You who believe in the doctrines of Copernicus assert that the earth revolves on an axis, which, as it sweeps round the sun, remains ever parallel to itself. This axis prolonged meets the celestial sphere in a point called the north pole. Now as the earth describes an orbit of nearly 200,000,000 miles in diameter, its axis prolonged will cut out of the sphere of the heavens a curve of equal dimensions, and the pole will appear to revolve and successively fill every point of this celestial curve in the course of the year. Now the north pole does not revolve in any such curve; it is ever fixed, and your theory is false. The Copernican could only reply, that all the premises were true, but that the conclusion was false. The pole of the heavens did revolve in just such a curve as stated, but such was the distance of the sphere of the fixed stars, that this curve of 200,000,000 miles in diameter was reduced to an invisible point!

Three hundred years have rolled away since this controversy began. The struggle has been long and arduous. The mind, baffled in one direction, has directed its energies in another: failing in one mode of research, it devises another, and thus struggling onward for three long centuries, it at length triumphs. The facts are developed, and the truth of the grand theory of Copernicus is vindicated and established, and the accuracy of these incredible statements is proved in the clearest manner.

As this discussion exhibits, clearly and beautifully, the progressive advances of human genius, I shall be pardoned for entering at some length, into an examination of the various attempts which have been made to resolve the problem of the parallax of the fixed stars. Indeed, the distance of the nearest fixed star is to become the unit of measure with which we are to traverse the innumerable worlds and systems by which we are surrounded, and on the accuracy with which it shall be determined will depend the correctness of the survey which we are soon to make.

Failing entirely in obtaining any parallactic angle with a base line of 8000 miles in length, the earth was employed to transport the observer from the first point of observation to a distance of 190,000,000 miles, there again to erect his telescope, and to send up his second visual ray to the far distant star, in the hope of finding a parallactic angle with a base of such enormous extent.

Permit me to illustrate the nature of this investigation. Suppose from the centre of a plane a solid granite rock, deep sunk and immoveable, rears its head far above the mists and impurities which float in the lower air. Ascending to the summit, the astronomer hews out some rough peak into the form of a vertical shaft. To this solid shaft he bolts the metallic plates which shall bear his telescope. The instrument is of a size and power commensurate with the grand objects which it is required to accomplish. Placed in a position such that its axis shall be exactly vertical, it is screw-bolted and iron-bound to the solid rock, with fastenings which shall hold it from year to year, fixed and immoveable as its rocky base.

To give more perfect precision to his work, the astronomer places in the focus of his eye-piece two delicate lines made from the spider's web, of a minuteness almost mathematical, which, by crossing at right angles, form a point of the utmost precision exactly in the axis of the telescope. These are in like manner fixed immoveably in their places, and now the machinery is prepared with which the observations are to be conducted.

Suppose the observations to commence to-night. On placing the eye to the telescope, and looking directly up to the zenith, a star enters the field of the instrument, and borne along by the diurnal motion of the heavens, advances towards the central point determined by the intersection of the spider's lines. In passing across the field of view, its minute diameter is exactly bisected by one of these delicate lines, and the exact moment, to the hundredth part of a second of time, is noted at which it passes the central point. This observation completed with all possible precision, in case no change in the apparent place of the star is produced by the revolution of the earth in its orbit, or by any other cause, on each successive night throughout the entire year the same phenomena will be repeated in the same precise order.

When the hour comes round, the star will enter the field, thread the spider's line, and reach the central point at the same precise instant, night after night, even for a thousand revolutions of the earth on its axis.

Such, then, is the delicate means employed in the examination of the problem of the parallax of the fixed stars; and nearly in this way did Bradley, the great English astronomer, prosecute this intricate investigation. If any change in the star's place is occasioned by the revolution of the earth in its orbit, sweeping, as it does, the spectator round the circumference of a track nearly 200,000,000 miles in diameter, it is easy to compute, not the amount, but the direction in which these changes will be accomplished. These computations were made by the astronomer, and all things being prepared, he commenced the series of observations which were to lead to the most important results. The discovery of absolute fixity in the star would be a great negative result, and any changes, no matter of what kind or character, could not fail to be detected.

Bradley.

Night after night was the astronomer found at his post, and as the months rolled slowly away, he began to perceive that his star, which, for a long time threaded the spider's line as it was in the act of passing the field of the telescope, began slowly to work off from this line, at last absolutely separating itself from it, and failing to reach the central point of the field at the precise instant first recorded. It soon became manifest that some cause or causes were operating to produce an apparent change in the place of the star; but what was the astonishment of Bradley to find that the changes in question could not be produced by parallax, for the motions detected were almost precisely opposite to those which would arise from this cause.

Long years of laborious examination were finally rewarded with two of the most brilliant discoveries ever accomplished by human skill and genius. The first of these demonstrated the fact that the sun and moon were so operating on the protuberant matter at the earth's equator, as to cause the axis of the earth to oscillate or revolve in a minute orbit, nodding to and fro under the influence of the configurations of these two controlling bodies, and following, in the most absolute manner, their relative positions. The effect of this variation, called *nutation*, is to cause all the stars to appear alternately to approach and recede from the pole. The real effect is to move the pole by the same amount.

The value of this change has been determined with the utmost precision, and although its entire effect does not shift the pole over a greater space than the fourth part of the apparent diameter of the planet Jupiter, its values, as deduced by different astronomers, and by different processes, scarcely differ by the fraction of a second of space. As a specimen of the accuracy attained in these delicate measurements, I will give three values recently obtained by the Russian astronomers. M. Busch, from Bradley's observations, obtains the value $9''.2320$; M. Liendhal, from observations at Dorpat, finds the value $9''.1361$; M. Peters, from right ascensions of Poralis, observed at Dorpat, fixes the value *Polaris* at $9''.2164$. The mean of the three values is $9''.2231$, the highest difference from which is less than the tenth part of one second of arc.

Valuable as was this discovery, it was actually surpassed by the importance of the second, for which we are in like manner indebted to Bradley. This second phenomenon consisted in an apparent movement of all the fixed stars in a minute orbit, which was accomplished in a year for every individual, and showed, in the most absolute manner, that it depended in some way on the orbitual revolution of the earth. For a long time, the true explanation of this phenomenon, which Bradley saw at once was not parallactic, eluded his highest sagacity. Potent thought and persevering reflection were, however, at last triumphant, and an explanation was finally reached, not only of the most satisfactory kind, but involving nothing less than an absolute demonstration of the orbitual motion of the earth, and a full confirmation of the velocity of light, whose prodigious swiftness had staggered the faith of many anxious to credit so marvellous a statement.

N

A few words will suffice to explain these phenomena. If we admit the progressive motion of light and the revolution of the earth in its orbit, it is manifest that the celestial bodies will not occupy in the heavens the places they appear to fill. Take, for example, the planet Jupiter, and even suppose the planet to be fixed. The telescope is directed to this object, and the light from the planet, streaming through the axis of the instrument, reaches the eye of the observer, and produces the visible image of the planet. But these very particles of light have occupied nearly forty minutes in passing from the planet to the eye of the observer. During these forty minutes, the earth has progressed in its orbit some 37,000 miles, and the spectator on the earth, borne along with it, must see the planet, not where it actually is, but where it was in appearance some forty minutes before. The same effect, in kind, is produced on the places of the fixed stars, and is called *aberration*. Understanding, now, that *some* effect must arise from these causes, (the velocity of light and the motion of the earth,) let us endeavour to render its nature clear, and the results palpable. To accomplish this, we must resort to the simplest means of elucidation.

Suppose a person were on the deck of a boat floating down the current of a river at any given rate per hour. As he moves steadily down the stream, he catches sight of an object on the shore, through which he proposes to send a rifle ball. The marksman will not aim directly at the object. Why? Because he knows that the rifle ball will partake of the boat's motion, and will be carried down, after it leaves the gun and before it reaches the mark, a distance equal to the progressive motion of the boat during the time of flight of the ball. To strike the mark he must therefore make this necessary allowance, and aim above it the required quantity. It is readily seen that the faster the boat moves, the greater will this allowance be.

Now reverse the proposition, and suppose a rifle fired on shore, and so directed as to fire a ball *down the barrel* of a gun on a moving boat. In case the two rifles are on the same exact level, and the axes of the barrels come precisely to coincide, it might be supposed, if the fixed one is fired at the exact instant the muzzles come precisely opposite to each other, that the ball from the one will pass down the other. But this, from a moment's reflection, is found to be false. The fixed rifle must be fired before the moving one comes opposite, and the allowance must be made by knowing how long the ball requires to move from the one gun to the other, and with what velocity the moving piece is descending. This computation being accurately made, the ball from the shore might be made to enter the muzzle of the moving rifle ; but while it is progressing down the barrel, the barrel itself is progressing down the stream, and hence, to avoid the pressure of the ball against the upper side of the barrel, we must fix it in an inclined position, and the bottom of the barrel must be as far up the stream as it will descend by the boat's motion during the progress of the ball down the barrel. Hence we see that the direction in which the barrel of the rifle which is to receive the ball is to be placed, is determined by the velocity of the ball, and the velocity of the boat which bears the rifle.

Now for the application. The particles of light coming from the fixed stars are the balls from the fixed rifle. The boat corresponds to the earth sweeping around in its orbit, and bearing with it the tube of the astronomer, down whose axis the particles of light must pass to reach the observer's eye. The velocity of the earth's motion is well known, and the amount by which the telescope must be inclined, to cause the light to enter, has been accurately determined, and from these two data the velocity of the light itself becomes known, and confirms, in the most satisfactory manner, the previously determined value of this incredible velocity, while the reality of the earth's motion is absolutely necessary to render the phenomena at all explicable.

Such were the beautiful results reached by Bradley ; and although nothing was gained with reference to the parallax, these preliminary discoveries were in themselves of the highest value, and prepared the way for subsequent observers, who with better means and more delicate instrumental aid, might prosecute the same great investigation.

Amid the numerous and diversified researches of Sir William Herschel, the problem of the parallax of the fixed stars could not fail to engage his attention, by its difficulty and importance. He devised a new means of prosecuting this research, which seemed to promise the most certain success. In his exploration of the heavens with his powerful telescopes, he discovered the curious fact that many fixed stars which to the unassisted eye appear as single objects, under the space-annihilating power of the telescope, are seen to consist of two, sometimes of three or more, individual stars, so close to each other that to the naked eye they blend into a single object.

Herschel, in the outset, conceived that this proximity of the stars was an accidental circumstance, and that where a pair could be found in which one individual was about double the other in magnitude, it might reasonably be inferred that the smaller of the two was *twice* as deeply sunk in space as the larger. If this hypothesis could be shown to be true, then would these objects present an admirable means of detecting with the greatest accuracy any change in their relative positions, occasioned by the habitual motion of the earth. In case their proximity was optical, or merely occasioned by the fact that the visual ray drawn to the one passed nearly through the other, it is manifest that, shifting greatly the position of the observer, the stars might be made to open out or close up on each other, or even the one revolve about the other. In employing this mode of investigation, the objects of comparison fell within the field of view of the same telescope, and almost all extraneous sources of error were eliminated.

Such was the plan devised, or rather perfected by Herschel (for his predecessors had already suggested it), and on the prosecution of which he entered with the zeal which ever distinguished this great astronomer. When he commenced his researches, some half-dozen double stars had been discovered and recorded. His first duty was to increase this number as rapidly as possible, and from his entire catalogue to select those best adapted to his purpose. Under his penetrating glance, the number of these curious duplex objects increased with astonishing rapidity, and he was himself startled with their frequent occurrence. However, with a mind fixed on his original design, he selected a large number of pairs, of such relative magnitudes, and in such positions, as promised the most certain success. Let it be remembered that many of these delicate objects were not divided from each other by a space greater than the thousandth part of the sun's diameter.

To ascertain the apparent change in the relative positions and distances of the stars composing these pairs, Herschel measured, with every care, the distance which separated them, and took the direction of the line drawn from the centre of the one to the centre of the other. Variations of distance and position occasioned by parallax, were easily computed in kind and character, and the great astronomer commenced and prosecuted his observations with sanguine hopes of success. One thing was certain; all parallactic movements would have a period of one year, since they arise from the annual revolution of the earth in its orbit; and at the end of this period the stars composing the double sets ought to return to the position occupied at the outset. What was Herschel's astonishment to find that, in many instances, the stars composing these pairs were actually in motion; but the movement was certainly not of a parallactic kind, for it neither agreed in direction nor in period with the effects of parallax. Here was another grand discovery! These double stars, which were scattered throughout the heavens with far greater profusion than accidental optical proximity could warrant, were found to be magnificent systems of revolving suns! They were united by the law of gravitation, and exhibited the wonderful spectacle of stupendous globes, moving in obedience to the same influences which hold the planets in their orbits, and guide the comets in their eccentric career. This is not the place to enter into detail concerning these wonderful objects.

While a new field of investigation, boundless and magnificent, was opened up to the human mind; while the great discoverer of these far-sweeping suns was more than rewarded for his toil and labour, the original object of his research was not only left unattained, but the method selected with so much reasonable hope of success, became utterly inapplicable. The parallactic and absolute motions of the systems of stars became so inextricably involved, that the imperfect micrometrical means of Herschel could not separate them.

Thus far, the efforts to obtain the distances of the stars had been unavailing. A negative solution had indeed been reached. That their distance was enormous, was made evident from the fact that the parallax had remained insensible even under the most careful and delicate instrumental tests. Any absolute solution began almost to be despaired of, when hope was again revived by the magnificent refracting telescopes, for which the world was indebted to the skill and genius of the celebrated Frauenhofer, of Munich. This great artist, aided by the profound science of Bessel, contrived and executed an instrument of extraordinary power, and especially adapted to the research for the parallax of the fixed stars.

Armed with a micrometrical apparatus of wonderful perfection, and capable of executing measures of great, as well as minute distances, the telescope was so arranged as to be carried forward by delicate machinery, with a velocity exactly equal to the diurnal motion of the object under examination. To give some idea of the delicacy of the contrivances with which these great telescopes have been provided, it is only necessary to state that the micrometer of the great Refractor of the Cincinnati observatory is capable of dividing an inch into 80,000 equal parts! When mechanical ingenuity failed to construct lines of mathematical minuteness, the spider lent his aid, and it is with his delicate web that these measures are accomplished. Two parallel spider's webs are adjusted in the focus of the eye-piece of the micrometer; and when the light of a small lamp is thrown upon them, the eye, on looking through the telescope, sees two minute golden wires, straight and beautiful, drawn across the centre of the field of view, and pictured on the heavens. These are within the control of the observer. He can increase or decrease their distance at pleasure, and so revolve them as to bring them into any position, every motion being accurately measured by properly divided scales.

Suppose, then, it is desired to take the distance and position of the stars forming a pair. The telescope is directed to them, and they are brought to the centre of the field of view. The clock-work is set in action; it takes up the ponderous instrument, weighing more than 2,500 pounds, and with the most astonishing accuracy it bears it onward, keeping its mighty eye fixed on the object under examination. The observer is thus left with both hands free to make his measures. He first revolves his micrometer spider's lines round until one of them shall exactly pass from centre to centre of the two stars. This position is noted, and from it is deduced the angle formed by this line with the meridian. He then revolves them a quarter of the circumference and they are then perpendicular to their former position. He now separates the wires until the one shall exactly bisect one star, while the other wire passes through the centre of the second star, reading this distance on the proper scale. He has fixed, in these two observations, the position and distance of the two components of the double set. Such is the precision attained in this work, that the most minute motions cannot escape detection. If the stars separate from each other at so slow a rate that a million of years would be required to perform the circuit of the heavens, their motion would be detected in half a year!

With machinery more delicate even than this, and better adapted to the purpose, and of a kind somewhat different, Bessel once more renewed the search after the unattainable parallax or the fixed stars. His great instrument, called the *heliometer*, was mounted as early as 1829, but a multitude of causes, and some unsuccessful efforts, delayed his principal operations up to August, 1837. Three great principles guided him in his selection of 61 in the Swan, as the star on which to perform his observations. First, it was affected by a very great *proper motion*, a characteristic which we will explain fully hereafter, and which indicated it to be among the nearest of all the stars. Secondly, its duplex character adapted it especially to the instrument he was about to employ. Thirdly, the region occupied by 61 Cygni contains a number of minute stellar points, close to the double star, and presenting admirable fixed points, to which the relative motions of the two components of the star to be measured might be referred.

With these advantages, and a magnificent instrument, Bessel commenced his observations. He measured the distance from the centre of the line joining the two stars, to two of the small stellar points, which served him as points of reference ; and this kind of observation was repeated night after night, whenever the stars were visible, from the middle of August, 1837, up to the end of September, 1838. The entire series of observations was then taken and corrected for every possible known error, and in case any appreciable change remained, it could only be attributed to parallax.

After a most careful and elaborate investigation, a variation commenced to show itself, increasing precisely as parallactic variation ought to increase, and diminishing as it ought to diminish. The period of these changes was precisely a year, and in all particulars there was an exact correspondence in kind with the changes which ought to be produced by parallax. But such was their minute character that Bessel hesitated.

During another year the observations were repeated. The same results came out, and the previous values were confirmed. A third year's observations, yielding precisely the same values, removed all doubt, and the great Koeningsburgh philosopher announced to the world that he had passed the impassable gulf of space, and had measured the distance to the sphere of the fixed stars ! But how shall I convey any adequate idea of this stupendous distance ? Millions and millions of miles serve only to confound the mind. Let us employ a different kind of unit.

Light, as we have seen, travels with a velocity of 12,000,000 miles in every minute of time. Hence, to reach us from the most remote of all the planets, Neptune, whose distance from th sun is about 3,000,000,000 miles, will require a journey of about four hours ; but to wing its flight across the interval which separates our sun from 61 Cygni, will require a period not to be reckoned by hours, or by days, or by months. Nearly ten years of time must roll away before its light, flying in every second 192,000 miles, can complete its mighty journey ! If the mind revolt at this conclusion, if the distance be too great for comprehension, if the scale of the universe thus suggested even stagger the imagination, I can only say that all subsequent observation has confirmed in the most satisfactory manner the accuracy of Bessel's results. This great astronomer first led the way across the mighty gulf which separates us from the fixed stars. The distance once passed, the route has become comparatively easy, and succeeding observers have determined the parallax of a sufficient number of stars to show that their results are entirely trust-worthy.

Having now succeeded in gaining a knowledge of the distance which separates our sun from its remote companions, we are prepared to extend our explorations of the universe. The question naturally arises, how are the stars distributed throughout space ? are they indifferently scattered in all directions, or are they grouped together into magnificent systems ? A cursory examination of the starry heavens with the naked eye shows us, that so far as the larger stars are concerned, they do not appear to have been distributed in the celestial sphere according to any determinate law ; but on applying the telescope, that luminous zone which under the name of the Milky Way, girdles the whole heavens, is found to be composed of minute stars, scattered like millions of diamond points on the deep blue ground of the sky.

Sir William Herschel conceived the idea, that it might be possible to fathom this mighty ocean of stars, and to determine its metes and bounds ; to give it figure, and to circumscribe its limits. It will not be difficult to explain, in a few words, the general outline of the plan adopted by this extraordinary man in the prosecution of this wonderful undertaking. In case we admit that the stars are of equal magnitudes, and at equal distances from each other, it would not be difficult to ascertain how far they extended in any given direction, the one behind the other. It is manifest, that in examining the heavens with a telescope of given power and aperture, we shall be able to count more stars in the field of view in those regions where they are so arranged as to reach farthest back into space ; and in case we know their absolute distance from each other, the number counted in any field of view will determine with accuracy the length of the visual ray reaching to the most remote star visible in that field.

Now, although the hypothesis that the stars are of equal magnitude, and are uniformly distributed through space, may not be rigorously true, yet doubtless the mean distances are not far from this hypothesis ; and although our results may only be approximate, yet as such they are to be relied upon, and they become the more interesting as they carry us to the utmost limits of human investigation. Armed with his mighty telescope, Sir William Herschel commenced the stupendous task of sounding the heavens, with the purpose of ascertaining whether the stars composing the Milky Way were unfathomable, or were bounded and circumscribed by definite limits.

Sweeping a circle round the heavens which cut this grand stratum of stars in a direction nearly perpendicular to its circumference, he directed his great telescope to a certain number of points along this circle, and as he moved slowly onward, counted all the stars visible in each field of view. It was fair to conclude, that wherever most stars were to be seen, there was the stratum deepest. Having gone entirely around the heavens, along the circumference of his circle, he had sounded the depth of the stars along a section of the Milky Way, and to obtain the figure of the section thus cut out was not a difficult matter.

He assumed a central point on paper to represent his point of observation. He then drew from this point lines radiating, and in the actual directions which he had given to his telescope while engaged in his explorations. On each of these indefinite lines he laid off a distance proportioned to the number of stars counted in the field of view in the direction which the line represented, and by joining these points thus determined, he formed a figure which represented the relative depths to which he had penetrated into space ; and in case he could be certain that he had gone absolutely through the stratum in every instance, and had grasped every star, even where the extent was most profound, the figures thus constructed would represent the form of the line cut from the outside boundary of the Milky Way by the plane of the circle in which the explorations had been made.

Did he then actually penetrate the deepest portions, or any portion of the Milky Way ? This was now his grand question, and to its decision he gave all his power and ingenuity. As a unit wherewith to measure the space-penetrating power of his telescopes, he assumed the power of the human eye, and knowing that stars of the sixth magnitude are within the reach of the unaided eye, he concluded, from the law regulating the decrease of light, that these minute stars were twelve times more distant than the nearest or brightest stars. Now, a telescope having an aperture such as to concentrate twice as much light as the eye, would penetrate into space twice as far, or would reach stars of the twenty-fourth order of distances, and so on for telescopes of all sizes. In this way, he concluded that his great forty-foot reflector, with a diameter of four feet, would penetrate 194 times as far as the naked eye, or that it would still see a star of the first magnitude, if it were carried backward into space, 2328 times its present distance !

Such, then, was the computed length of the *sounding-line* employed in gauging these mighty depths. Suppose, then, it was required to determine whether this line actually penetrated any given region of the Milky Way. Even with a single telescope, a series of experiments may be performed which go very far to determine this great question. As the space-penetrating power of a telescope depends on the diameter of its aperture, it is easy to give to the same instrument different powers, by covering up, by circular coverings, certain portions of its object-glass. Take circles of paste-board, or any other suitable material, and, in the first, cut an opening one inch in diameter, in the second an opening of two inches, and so on, up to the diameter of the object-glass. These diaphragms being successively applied to the object-glass, give to the telescope space-penetrating powers proportioned to the diameter of the openings.

In this way, Herschel prepared himself to explore one of the

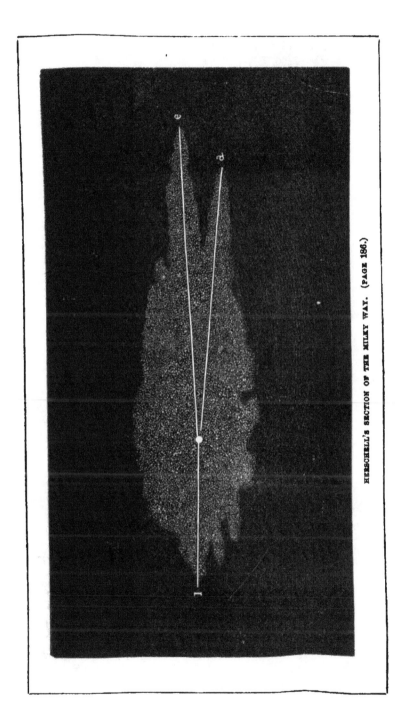

HERSCHELL'S SECTION OF THE MILKY WAY. (PAGE 186.)

deepest portions of the Milky Way. The spot selected was a nebulous or hazy cloud in the sword-handle of Perseus, in which, to the naked eye, not a solitary star was visible. I have many times examined the same object, which is certainly one of the most magnificent the eye ever beheld. With the lowest telescopic aid, many stars are rendered visible, surrounded by a hazy light, in which minute glimpse points are occasionally to be seen. As the space-penetrating power was increased, the bright spots of light were successively resolved into groups of brilliant stars, and more nebulous haze came up from the deep distance, indicating that the visual ray was not long enough to fathom the mighty distance. At last the full power of his grand instrument was brought to bear, when a countless multitude of magnificent orbs burst on the sight, like so many sparkling diamonds on the deep blue of the heavens. There was no haze behind; the telescopic ray had shot entirely through the mighty distance, and the clear deep heavens formed the back-ground of the brilliant picture.

Thus did Herschel penetrate to the limits of the Milky Way, and send his almost illimitable sounding line far beyond, into the vast abyss of space, boundless and unfathomable. And now do you inquire the depth of this stupendous stratum of stars? The answer may be given, since we have the unit of measure in the distance of stars of the first magnitude. Light, with its amazing velocity, requires ten years to come to us from the nearest fixed stars, and yet Sir William Herschel concluded, from the examinations he had been able to make, that in some places the depth of the Milky Way was such that no less than five hundred stars were ranged one behind the other in a line, each separated from the other by a distance equal to that which divides our sun from the nearest fixed star. So that, for light to sweep across the diameter of this vast congeries of stars, would require a period of a thousand years at the rate of 12,000,000 miles in every minute of time !

The countless millions of stars composing the Milky Way appear to be arranged in the form of a flat zone or ring, or rather stratum, of irregular shape, which I shall explain more fully hereafter. Its extent is so great as properly to form a universe of itself. If it were possible, to-night, to wing our flight to any one of the bright stars which blaze around us, sweeping away from our own system, until planet after planet

fades in the distance, and finally the sun itself shrinks into a
mere star, alighting on a strange world, that circles round a new
and magnificent sun, which has grown and expanded in our
sight, until it blazes with a magnificence equal to that of our
own, here let us pause and look out upon the starry heavens
which now surround us.

We have passed over sixty millions of millions of miles. We
have reached a new system of worlds revolving about another
sun, and from this remote point we have a right to expect a new
heaven, as well as a new earth on which we stand. But, no.
Lift up your eyes, and lo ! the old familiar constellations are all
there. Yonder blazes Orion, with his rich and gorgeous belt ;
there comes Arcturus, and, yonder, the Northern Bear circles
his ceaseless journey round the pole. All is unchanged, and the
mighty distance over which we have passed is but the thousandth
part of the entire diameter of this grand cluster of suns and
systems ; and although we have swept from our sun to the
nearest fixed star, and have travelled a distance which light
itself cannot traverse in less than ten years, yet the change
wrought by this mighty journey, in the appearance of the heavens,
is no greater than would be produced in the relative positions of
the persons composing this audience to a person near its centre,
who should change his seat with his immediate neighbour !

Such, then, is the scale on which the starry heavens are built.
If, in examining the magnificent orbits of the remoter planets,
and, in tracing the interminable career of some of the far-
sweeping comets, we feared there might not be room for the
accomplishment of their vast orbits, our fears are now at an end.
There is no jostling here ; there is no interference, no perturbation of the planets of one system by the suns of another. Each
is isolated and independent, filling the region of space assigned,
and, within its own limits, holding on its appointed movements.

Thus far we have spoken only of the Milky Way. In case it
be possible to pierce its boundaries, and pass through into the
regions of space which lie beyond, the inquiry arises, What meets
the vision there ? What lies beyond these mighty limits ? Does
creation cease with this one great cluster, and is all blank beyond
its boundary ?

Here, again, the telescope has given us an answer. When we
shall have travelled outward from our own sun, and passed, in a
straight line from star to star, until we shall have left behind us,

in grand perspective, a series of five hundred suns, we then stand on the confines of our own great cluster of stars. All behind blazes with the light of countless orbs, scattered in wild magnificence, while all before us is deep, impenetrable, unbroken darkness. No glance of human vision can pierce the dark profound.

But summoning the telescope to our aid, let us pursue our mighty journey through space; far in the distance we are just able to discover a faint haze of light, a minute luminous cloud which comes up to meet us, and towards this object we will urge our flight. We leave the shining millions of our own great cluster far behind. Its stars are shrinking and fading; its dimensions are contracting. It once filled the whole heavens, and now its myriads of blazing orbs could almost be grasped with a single hand. But now look forward. A new universe, of astonishing grandeur, bursts on the sight. The cloud of light has swelled and expanded, and its millions of suns now fill the whole heavens.

We have reached the clustering of millions of stars. Look to the right—there is no limit; look to the left—there is no end. Above, below, sun rises upon sun, and system on system, in endless and immeasurable perspective. Here is a new universe, as magnificent, as glorious as our own; a new Milky Way, whose vast diameter the flashing light would not cross in a thousand years. Nor is this a solitary object. Go out on a clear cold winter night, and reckon the stars which strew the heavens, and count their number; and for every single orb thus visible to the naked eye the telescope reveals a *universe*, far sunk in the depths of space, and scattered with vast profusion over the entire surface of the heavens.

Some of these blaze with countless stars, while others, occupying the confines of visible space, but dimly stain the blue of the sky, just perceptible with the most powerful means that man can summon to the aid of his vision. These objects are called clusters and nebulæ; clusters, when near enough to permit their individual stars to be shown by the telescope—nebulæ, when the mingled light of all their suns and systems can only be seen as a hazy cloud.

Thus have we risen in the orders of creation. We commenced with a planet and its satellite; we rose to the sun and its revolving planets, a magnificent system of orbs, all united into one great family, and governed by the same great law; and we now find millions of these suns clustered and associated together in the formation of distinct universes, whose number, already revealed to the eye of man, is not to be counted by scores or hundreds, but has risen to thousands, while every increase of telescopic power is adding by hundreds to their catalogue.

Let us now explain these "island universes," as the Germans have aptly termed them, and attempt approximately to circumscribe their limits, and measure their distances from us, and from each other. Sir William Herschel, to whom we are indebted for this department of astronomy, conceived a plan by which it was possible approximately to sound the depths of space, and determine, within certain limits, the distance and magnitudes of the clusters and nebulæ within the reach of his telescopes. To convey some idea of his method of conducting these most wonderful researches, imagine a level plane, of indefinite extent, and along a straight line, separated by intervals of one mile each; let posts be placed, bearing boards on which certain words are printed in letters of the same size. The words printed on the nearest board, we will suppose, can just be read with the naked eye. To read those on the second, telescopic aid is required; and that power which suffices to enable the letters to be distinctly seen, is exactly double that of the unaided eye. The telescope revealing the letters at the distance of three miles is threefold more powerful than the eye, and so of all the others. In this way we can provide ourselves with instruments whose space-penetrating power, compared with that of the eye, can be readily obtained.

Now to apply these principles to the sounding of the heavens. The eye, without assistance, would follow and still perceive the bright star Sirius, if removed back to twelve times its present distance. After this, as it recedes, it must be followed by the telescope. Suppose, then, a nebula is discovered with a telescope of low power, and it is required to determine its character and distance. The astronomer applies one power after another, until he finally employs a telescope of sufficient reach to reveal the separate stars of which the object is composed, which shows it to be a cluster; and since the space-penetrating power of this instrument is known, relative to that of the human eye, in case the power is one hundred times greater than that of the eye, then would the cluster be located in space one hundred times farther than the eye can reach, or twelve hundred times more remote than Sirius, or at such a distance that its light would only reach our earth after a journey of 120,000 years !

Such was Herschel's method of locating these objects in space. Some are so remote as to be far beyond the reach of the most powerful instruments, and no telescopic aid can show them other than nebulous clouds of greater or less extent. It was while pursuing these grand investigations that Herschel was led to the conclusion, that among the nebulæ which were visible in the heavens, there were some composed of chaotic matter, a hazy, luminous fluid, like that occasionally thrown out from comets on their approach to the sun.

Among these chaotic masses he discovered some in which the evidences of condensation appeared manifest, while in others he found a circular disc of light, with a bright nucleus in the centre. Proceeding yet farther, he found well formed stars surrounded by a misty halo, which presented all the characteristics of what he now conceived to be nebulous fluid. Some of the unformed nebulæ were of enormous extent, and among those partially condensed, such as the nebulæ with planetary discs, many were found so vast that their magnitude would fill the space occupied by the sun and all its planets, forming a sphere with a diameter of more than 6,000,000,000 miles. Uniting these and many other facts, the great astronomer was finally brought to believe that worlds and systems of worlds might yet be in process of formation, by the gradual condensation of this nebulous fluid, and that from this chaotic matter originally came the sun and all the fixed stars which crowd the heavens. This theory, extended, but not modified, in the hands of Laplace, is made to account for nearly all the phenomena of the solar system, and has been already referred to in a former lecture.

For a long time, this bold and sublime speculation was looked upon, even by the wisest philosophers, with remarkable favour. The resolution of one or two nebulæ (so classed by Herschel), with the fifty-two feet reflector of Lord Rosse, has induced some persons to abandon the theory, and to attempt to prove its utter impossibility. All that I have to say, is, that Herschel only adopted the theory after he had resolved many hundreds of nebulæ into stars ; and if there ever existed a reason for accepting the truth of this remarkable speculation, that reason has been scarcely in any degree affected by recent discoveries.

I have examined a large number of these mysterious objects, floating on the deep ocean of space like the faintest filmy clouds of light. No power, however great, of the telescope, can accomplish the slightest change in their appearance. So distant that their light employs (in case they be clusters) hundreds of thousands of years in reaching the eye that gazes upon them, and so extensive, even when viewed from such a distance, as to fill the entire field of view of the telescope many times. Sirius, the brightest, and probably the largest, of all the fixed stars, with a diameter of more than a million of miles, and a distance of only a single unit, compared with the tens of thousands which divide us from some of the nebulæ; and yet this vast globe, at this comparatively short distance, is an inappreciable point in the

Lord Rosse.

field of the telescope. What, then, must be the dimensions of those objects, which, at so vast a distance, fill the entire field of view even many times repeated?

Herschel computes that the power of his great reflector would follow one of the large clusters if it were plunged so deep in space that its light would require 350,000 years to reach us, and the great telescope of Lord Rosse would pursue the same object probably to ten times this enormous distance.

Such examinations absolutely overwhélm the mind, and the wild dream of the German poet becomes a sort of sublime reality :—

" God called up from dreams a man into the vestibule of Heáven, saying, ' Come thou hither, and see the glory of my house.' And to the servants that stood around his throne he said, ' Take him, and undress him from his robes of flesh : cleanse his vision, and put a new breath into his nostrils ; only touch not with any change his human heart—the heart that weeps and trembles.' It was done : and, with a mighty angel for his guide, the man stood ready for his infinite voyage ; and from the terraces of Heaven, without sound or farewell, at once they wheeled away into endless space. Sometimes with the solemn flight of angel wing they fled through Zaarahs of darkness, through wildernesses of death, that divided the worlds of life ; sometimes they swept over frontiers, that were quickening under prophetic motions from God. Then, from a distance that is counted only in Heaven, light dawned for a time through a sleepy film ; by unutterable pace the light swept to *them ;* they, by unutterable pace, to the light. In a moment, the rushing of planets was upon them ; in a moment, the blazing of suns was around them.

" Then came eternities of twilight, that revealed, but were not revealed. On the right hand and on the left, towered mighty constellations, that, by self-repetitions and answers from afar, that, by counter-positions, built up triumphal gates, whose architraves, whose archways, horizontal, upright, rested, rose, at altitude by spans that seemed ghostly from infinitude. Without measure were the architraves, past number were the archways, beyond memory the gates. Within were stairs that scaled the eternities below ; above was below, below was above, to the man stripped of gravitating body : depth was swallowed up in height insurmountable—height was swallowed up in depth unfathomable. Suddenly, as thus they rode from infinite to infinite, —suddenly, as thus they tilted over abysmal worlds, a mighty cry arose, that systems more mysterious, that worlds more billowy, other heights and other depths, were coming, were nearing, were at hand.

" Then the man sighed, and stopped, shuddered, and wept. His overladened heart uttered itself in tears ; and he said ' Angel, I will go no farther ; for the spirit of man acheth with

o

this infinity. Insufferable is the glory of God. Let me lie down in the grave and hide me from the persecution of the infinite ; for end, I see, there is none.' And from all the listening stars that shone around issued a choral voice, ' The man speaks truly : end there is none, that ever yet we heard of ! ' ' End is there none ? ' the angel solemnly demanded : ' Is there indeed no end ? and is this the sorrow that kills you ? ' But no voice answered, that he might answer himself. Then the angel threw up his glorious hands to the Heaven of Heavens, saying, ' End is there none to the universe of God. Lo ! also there is no beginning.' "

Relative Telescopic appearance of the Planets.

LECTURE X.

AVING reached, in the course of the preceding lecture, to the outermost confines of the visible creation, let us now return home from this survey of the "island universes" which crowd the illimitable regions of space, to the stars which compose our own cluster, and learn how far the human mind has progressed in its examination of the millions of suns, which constitute, in a more definite sense, our own Milky Way.

We have already seen that the parallax of 61 Cygni rewarded the laborious and extraordinary efforts of Bessel. The example set by this great astronomer encouraged those who followed him; and while his results in this particular case have been confirmed in the most astonishing manner, the distances of many other stars have been obtained, until a sufficient amount of data has been accumulated to determine the approximate distances of the spheres of the fixed stars of different magnitudes. Struve estimates the mean distance of stars of the first magnitude to be 986,000 times the radius of the earth's orbit, or so remote, that their light reaches us only after a journey of fifteen years and a half. Stars of the second magnitude send us their light in twenty-eight years, those of the third magnitude in forty-three years; while the light from

o 2

stars of the ninth magnitude only reaches the eye of the observer after traversing space for five hundred and eighty-six years, at the rate of 12,000,000 miles in every minute of time.

My range of investigation does not permit me to explain at this time how these extraordinary conclusions have been reached. The reasoning, however, is close and clear, and the results are no doubt approximately correct.

Such, then, are the distances separating man from the objects of his research. To have attained to a knowledge of these distances even, is sufficiently wonderful, but what we are about to reveal as the results of human investigation among these far distant orbs, cannot fail to fill the mind with astonishment, and demonstrate the great truth, that " man has been made but a little lower than the angels."

Before it became possible to examine with absolute certainty the places of the stars, with a view to ascertain their absolute fixity, many difficult preliminary preparations had to be accomplished. Instruments of the most perfect kind must be provided, not only in their optical performances, but in their space-dividing machinery. Moreover, the places of the stars, as determined by the best telescopes, must be corrected for every possible instrumental error. The two points to which the stars referred are the north pole and the vernal equinox. In case any motions belong to these points, their amounts and directions must be ascertained and allowed for. Then the effects of refraction, and of the aberration of light, were indispensable to a perfect investigation of the absolute places of the stars.

All these, and many other preliminary matters, having been satisfactorily determined, it became possible to examine, in the most critical manner, the places of the stars, and to learn whether, indeed (as had been supposed for thousands of years), their configurations were eternal and unchangeable, or whether they moved among themselves with a motion rendered so slow by their immense distance, as hitherto to have escaped the most scrutinising watch.

Fully armed with the necessary instruments, it did not require many years to determine the grand truth, that among the tens of thousands of stars which fill the heavens, not a solitary one, in all probability, is in a state of absolute rest. Many were found to move so swiftly, that their velocity was determined even in a single year ; while others, in consequence of their enormous distance, may require centuries to detect any appreciable change. In the outset these extraordinary movements seemed to be directed by no law; some stars were sweeping in one direction, and some in another. Motion, ceaseless, eternal motion, seems to be stamped on the entire universe; and while the stars are pursuing their mighty orbits, we cannot resist the idea that our own sun, the centre of our great planetary system, itself a star, must participate in the general movement, and is, in all probability, urging its flight, accompanied by all its planets, satellites, and comets, to some unknown region of space.

The revolution of the stars, the organisation of the grand cluster with which our sun is associated, the demonstration of the sun's absolute translation through space, its direction, velocity, and period, are the topics to which I invite your attention in the closing lecture of the present course.

When forced to acknowledge the rotation of our globe on its axis, and its swift orbitual motion, surrounded by wheeling planets and flying comets, the mind naturally retreats to the sun as the great immoveable centre, where it can rest and contemplate these circling worlds. But even here, as we shall presently see, there is no rest. The sun himself becomes a subordinate member of a grander combination of worlds, and, obedient to higher influence, sweeps around in its unmeasured orbit.

We shall present a rapid summary of the evidences of change among the fixed stars, and then proceed to develope the reasoning by which the direction and velocity of the sun's motion in space has been determined.

More than two thousand years ago, the celebrated Greek astronomer, Hipparchus, was astonished by the sudden bursting forth of a brilliant star in a region of the heavens, where none had previously existed. Up to this time, no doubt of the immutability of the starry sphere seems to have been entertained; and while the philosopher gazed and wondered, he resolved to execute a work, from which posterity might learn the changes of the celestial sphere. He undertook and completed his great catalogue of the places of a thousand stars, locating them with all the accuracy permitted by the rude instruments then in use. Subsequent observers, by comparing their own determined positions of the stars with their places as fixed on the catalogue of Hipparchus, could readily perceive any sensible change which might occur in their configuration, the appearance of new stars, or the disappearance of those which had once existed.

The sudden breaking forth of a new star is a phenomenon of such wonderful character, that we might well doubt the possibility of its occurrence, if we were obliged to rely on the historical account transmitted to us from the time of Hipparchus. But, fortunately, more than one brilliant example of the kind has occurred in modern times, presenting the most unequivocal evidence of the reality of this inexplicable wonder.

In 1572, a new star of great splendour appeared suddenly in the constellation Cassiopeia, occupying a position which had previously been blank. This star was first perceived by Schuler of Wittemberg, on the 6th of August. It was detected by

Tycho Brahe.

Tycho, the Danish astronomer, on the 11th of the following November, and the wonder produced by this most extraordinary phenomenon induced him to give to the star the most unremitting attention. Its magnitude increased until it is said to have surpassed even Jupiter in splendour, and finally became visible in the day time. It retained its greatest magnitude but for a very short time, when it commenced to diminish in brilliancy, changing from white to yellow, then to reddish, and finally it became faintly blue ; and so, diminishing by degrees, it vanished from the sight in March, 1574, and has never since been seen.

In the year 1604, while the scholars of Kepler were engaged in observations of Mars, Jupiter, and Saturn, then in close proximity to each other, having been interrupted a day or two by clouds, on the return of fine weather, Maestlin was astonished to find near the planets then in the constellation Ophiuchus, a brilliant star, which certainly had not been there a few days before. This object attracted the attention of all the great

astronomers then living, and was particularly observed by
Galileo and Kepler. It is said to have attained a splendour
equal to that of the planet Venus, and from this, its greatest
brilliancy, it gradually declined, until, about the beginning of
1606, it ceased to be visible, and no telescopic power has since
been able to detect any star in the place once occupied by this
remarkable stranger.

Although observed with the greatest care, no sensible parallax
was ever detected in either of these objects, and no doubt exists
as to their occupying the region of the fixed stars. Many other

Tycho Brahe's house and observatory.

less remarkable examples are on record, but up to the present
no satisfactory explanation of this remarkable phenomenon has
been given. Whether it indicates the actual destruction of some
magnificent system, or the revolution of these stars in orbits of
great eccentricity, causing them to appear to us, like the comets,
only in the perihelion points of their mighty orbits, is equally
uncertain. One thing is certain: they present evidence of

change in the starry heavens of the most startling and irresistible character.

While new stars have occasionally made their appearance, to astonish mankind with their brilliancy, there are many well authenticated cases of the entire disappearance of old stars, whose places had been fixed with a degree of certainty not to be doubted. In October, 1781, Sir William Herschel observed a star, No. 55 in Flamsted's catalogue, in the constellation

Flamsted.

Hercules. In 1790, the same star was observed by the same astronomer, but since that time no search has been able to detect it. The star is gone; whether never to return, it is impossible to say. A like disappearance has occurred with reference to the stars numbered 80 and 81, both of the fourth magnitude, in the same constellation. In May, 1828, Sir John Herschel missed the star numbered 42 in the constellation Virgo, which has never since been seen. Examples might be multiplied, but it is unnecessary.

In these cases the stars have been lost entirely; no return has ever been marked; and but for the discovery of another class of remarkable objects among the stars, no return would probably ever have been suspected. If I could direct your attention to-night to a brilliant star named Algol, in the head of Medusa, and bring a powerful telescope to aid in your examinations, this star, while you are watching it, might be seen to lose its splendour, and from its rank of the second magnitude, to decline in brightness, until it would scarcely be visible to the naked eye. Having reached a certain limit, it would commence and increase, and by slow degrees assume its original splendour. This decrease and increase is actually accomplished in about eight hours. Having regained its usual light, it remains stationary for about two days and a half, and then repeats the changes already detailed; and thus have its periodical fluctuations continued since the date of its discovery, with the most astonishing regularity. The bright star marked Beta, in the constellation Lyra, is known to pass from the third to the fifth magnitude, and to regain its light in a period of six days and nine hours. These are called *periodical* stars, and a sufficient number have already been detected to present a progressively increasing series of periods from two days twenty hours, up to four hundred and ninety-four days, and in one case even many years.

Here, again, are phenomena, indicative of extraordinary activity in these remote regions of space. No explanation of these changes has yet been given in all respects satisfactory. Some have attributed them to the existence of dark spots on the stars, which, by rotation on an axis, periodically present themselves, and thus dim the lustre of the stars. Others think the changes are due to the revolution of large planets about the stars, which, by coming between the eye and the star, eclipse a portion of its light ; while a third class conceive the fluctuations to arise, in some instances at least, from an orbital motion of the stars in orbits of excessive elongation, and so located as to have their greater axis directed towards our system.

It will be seen that this theory may be readily extended so as to embrace the new stars already referred to, and even to account for those which had been lost from their places in the heavens. Here, however, we enter the confines of the uncertain. Centuries may roll away before the true explanation of these astonishing changes shall be given ; but the mind is on the track, and with a steady and resistless movement, is slowly pushing its investigations deeper and still deeper into the dark unknown.

While the phenomena of the new and lost stars, and the fluctuations in the light of the variable ones, give undeniable evidence of constant change in what Aristotle was pleased to call the eternal and incorruptible heavens, Herschel's brilliant discovery of the orbital motion of the double stars gave to the mind the opportunity of determining the nature of the law which sways the movements in these distant regions of space. It was natural, in the first efforts to compute the orbits of the double stars, to adopt the hypothesis that they attracted each other by the same law which prevails in the planetary system. Results did not disappoint expectation. Gravitation, which Newton, in the outset of his great discovery, had boldly affirmed exerted its influence wherever matter existed or motion reigned, was extended, in the most absolute manner, to the region of the fixed stars. There, at a distance from our own system almost inconceivable, suns, and systems of suns, rising in orders ot greater complexity, revolving with swift velocity, or with slow and majestic motion, bore testimony, ample and unequivocal, to the truth of the great law of universal gravitation.

Every particle of matter in the universe attracts every other particle of matter, with a force which is proportioned directly to the mass, and which decreases as the square of the distance at which it operates increases. This is no longer a bold hypothesis. The double star marked Zeta, in the constellation Hercules, has been subjected to the analysis of the computer. The elements of its orbit have been obtained, and, true to its predicted period, it has actually performed an entire revolution in a period of thirty-five years. The components of the star Eta, in the Northern Crown, revolve around their common centre in about forty-four years. Both of these pairs have completed an entire revolution since their discovery. Many others might be named, but my only object at present is to exhibit the evidence which shall remove all doubt as to the actual extension of the law of gravitation to the fixed stars.

Let it be remembered that this department of astronomy is yet in its infancy. Thousands of double stars have been detected, and every year adds hundreds to the list. Among these, a large proportion must prove to be binary systems, varying in their periods of revolution, from thirty years or less, up to many thousands, perhaps millions of years.

The association of two suns naturally suggests the possible union of a greater number, forming more complicated systems. This idea has been verified ; a large number of triple systems has been discovered. In a few instances, quadruple sets have been found, of which a remarkable example exists in the constellation of the Harp. Here were found four suns, arranged in pairs of two. The components of the first pair revolve around each other in about one thousand years ; those of the second pair appear to require about double that period, while one pair revolves about the other in a period which, determined roughly from their distance, cannot fall much below a million of years ! The evidence of the physical union of these four stars into one grand system rests, at present, on the ascertained fact that their proper motions are the same.

From quadruple systems we rise, by analogy, still higher, until we find hundreds, sometimes thousands, of stars compacted together in so small a compass that their proximity cannot be the effect of accident. Look at the beautiful little cluster called the Pleiades ; an ordinary eye may there see six or seven stars. One of very great power has been known to count fourteen in this group, while the telescope increases the number to hundreds ; and yet the space in which they are located might easily be covered by the moon.

Suppose an indifferent scattering of the stars through space, and compute the chances that such a number would fall together at any one point, and we shall find not one chance out of millions in favour of such an accident. We are, therefore, forced to the conclusion that here is a more magnificent order,—one in which hundreds of suns, surrounded by their subordinate worlds, are all united by gravitation into one grand system. This is not a solitary example. Many of these beautiful objects, comparatively close to our sun, are found in the heavens, leading the mind gradually up to the contemplation and examination of that mighty system of systems, that great cluster of clusters, the Milky Way, of which all these are but subordinate groupings, vast in themselves, but when compared with the whole, mere units among the millions of which it is composed.

From what we have seen, it is impossible to avoid the conclusion that gravitation exerts its power among the myriads of shining orbs which strow the Milky Way. The innumerable suns which form this stupendous cluster must feel the reciprocal influence of each other ; and nothing short of the centrifugal force arising from orbital motion can balance this universal attractive power, and give to this grand system the great characteristic of stability.

Herschel succeeded, at least approximately, in sounding the profundities of the Milky Way, and fixed the relative position of our own sun among the stars by which it is surrounded. He found it to be located not very distant from the centre of the great stratum, and near the great line where the principal current of stars divides into two great streams, which for a time separate from each other, but finally reunite in a distant region of the heavens.

Having accomplished thus much, this great astronomer attempted the resolution of the grand problem of the sun's movement through space. This investigation is so lofty, so daring, and utterly incomprehensible at the first glance, that its mere announcement produces little effect on the mind.

Consider, for one moment, what it involves. Man is located on a planet almost infinitely larger than himself. This planet is swiftly revolving on its axis, and in its orbit round a great central luminary, the sun. The daring philosopher participates in all these motions. He provides himself with instruments which measure the distances and positions of the almost infinitely distant fixed stars. These fixed stars, when subjected to his critical examination, cease to be fixed, and are found to be moving with astonishing velocity in all directions. Among these he numbers his own sun, and, although borne along in the progressive motion of his own great centre, he ventures to attempt the determination of the fact of its actual motion, the direction in which it moves, and the velocity with which it is sweeping through space.

This problem is so wonderful, that I beg your earnest attention while the effort is made to simplify the reasoning by which its resolution has been accomplished.

Before the actual motions of the earth were discovered, the sun, moon, and planets, as well as the stars, *appeared* to move in certain directions, and with certain velocities, not easily explained. The rotation of the earth on its axis rendered a clear explanation of the diurnal movements of the heavenly bodies, and its orbital motion around the sun explained the sun's apparent annual movement among the fixed stars. Thus it is seen and readily apprehended, that in case the spectator is progressing, his actual motion may be transferred to distant bodies under examination, and these may appear to move while he seems to be at rest.

Now, in case the sun is sweeping towards any quarter of the heavens, it must carry with it all its planets, satellites, and comets. The earth is borne along in common with its companions, and the observer on its surface will transfer his own movement through space to the distant objects which only appear to change their places, in consequence of his own translation through space. Thus the distant stars may be affected with a parallactic change, not to be confounded with that produced by the revolution of the earth in its orbit, but occasioned by the fact, that while the earth revolves around the sun, she is carried forward by this luminary in his journey through space. As the whole system participates in this motion, in case the planets are inhabited, their astronomers will detect in the fixed stars the parallactic motion due to the sun's movement; and hence this change among the stars may be properly termed their *systematic* parallax.

Herschel commenced his examination of this great problem by forming a catalogue of stars situated in all parts of the heavens, in which an appreciable amount of *proper motion* had been detected and measured. Now, in case this apparent motion of the stars could be attributed to the movement of the solar system through space, a close scrutiny of the directions in which the stars appeared to move would indicate the direction in which the observer, carried along with the sun, was passing through space.

In case a person is travelling on a railway, in a direct line through a forest of trees, as he advances, all objects towards which he is moving will appear to open out or separate from each other, while those left behind will appear to close up. If, then, the astronomer, borne along by the movement of the sun through the vast *forest* of stars by which he is surrounded, desires to ascertain the direction in which he is progressing, let him search the heavens until he finds a point where the stars seem to be increasing their distance from each other. Should he find such a point, let him confirm his suspicions by looking in the direction precisely opposite and behind him, and in case he finds the stars located in this region closing up on each other, he may fairly conclude that he has found the direction in which he is moving, and a rigid coincidence of all the phenomena would demonstrate the accuracy of his conclusions.

Such was the general train of investigation adopted by Herschel. After as extended an examination as the data with which he was then furnished permitted, he announced his belief that a part of the proper motion of the fixed stars must be attributed to the effect of systematic parallax, and that the solar system was moving through space towards a point in the constellation Hercules.

The announcement of this astonishing result was received with hesitation and doubt by the best living astronomers, and Herschel died before any confirmation of his great theory had been obtained. After his death, for nearly half a century, no mind seemed willing to renew the investigation. The theory fell into disrepute, and was only regarded as a bold and sublime speculation, but not founded on any well-determined observations.

Within a few years, the problem has engaged the attention of the distinguished astronomers of Russia. Argelander, of Bonn, led the way, and, by a train of reasoning based upon extensive and accurate observations, has sustained and demonstrated, in the most undeniable manner, not only the general truth of Herschel's theory, but has even confirmed the direction in which that astronomer believed the solar system to be moving.

Here again permit me to attempt a popular explanation of Argelander's reasoning. Suppose a single star to have its place fixed absolutely by observation on the first day of the year 1700. One hundred years after, its place is again determined, when it is found to have shifted its position. Conceive the star to have so moved as to reach the meridian earlier than it formerly did. When on the meridian, its old place will be behind or east of the new place, and a line joining the old and new places will show the direction in which the star has been moving; and the distance between the two places will exhibit the amount of motion in one hundred years. If the star do not move exactly north or south, its line of direction will form an angle with the meridian, whose value is determined from a comparison of the old and new places of the star.

Argelander commenced by selecting five hundred stars, in all regions of the heavens, whose places had been well determined by preceding astronomers. The preference was of course given to those which had been longest subjected to observation. Having himself determined the new places of all these stars, a comparison of his own with previously observed positions determined the direction in which these stars were moving, and their rates of motion. The angles formed by the lines along which each star was progressing, with the meridian, became known from observation; and these angles we shall call the *observed angles of direction*. Now, it is not difficult to compute the directions in which the stars would appear to move, if their motion be produced by the movement of the solar system.

Suppose, for example, that the sun, with its planets, is sweeping exactly towards the north pole of the heavens, then would all the stars appear to move towards the south. Those in the equator would move with the swiftest velocity from the north pole; but those nearest the pole would appear to separate from each other, while their recess from the pole would be comparatively slight. To render this reasoning still plainer, imagine this room to be pierced on every side, so that an eye placed at the centre could see every star in the heavens through the openings. Through each of these holes conceive iron rods to pass, all meeting at a point in the centre, and all directed exactly to the stars. On the outside let golden balls be fixed to the extremities of these rods, to represent the stars. Now, grasping the extremities of all these rods in the hand, urge the point where they all unite towards the north pole, and watch the movement of the balls at the outer extremities of the rods. The ball corresponding to the north star will scarcely seem to move, because the eye travels directly towards it. The balls corresponding to the stars on the equator, having their rods perpendicular to the direction of the motion of the central point, will sweep swiftly towards the south. The idea once gained, there is no difficulty in its application.

The visual rays drawn to the stars correspond to the rods, and these rays, meeting in the eye of the observer, are carried forward by the sun in its progression through space. I have supposed the system to move due north ; but in case the motion be assumed in any other direction, it is easy to compute the changes consequent. Understanding these preliminary statements, we are prepared to follow Argelander in his investigation.

The five hundred stars selected for examination were divided into three groups, according to the amount of annual proper motion. The first contained only such stars as were seen to move with a velocity not less than one second of space in a year. Although this motion may appear excessively slow, yet its direction in one hundred years may be determined with very great precision. A general examination of the direction in which the stars of this first group appeared to move, indicated the quarter of the heavens towards which the solar system must be progressing ; and now commenced the investigation, having for its object the discovery of the exact point. To accomplish this, a point was assumed, and on the hypothesis that it was correctly chosen, the directions of the motion of all the stars composing the first group were computed, and the angles formed by their lines of direction with the meridian were determined.

If the motion of these stars was the effect of systematic parallax, and if the direction of the solar movement had been

accurately chosen, then would the *computed angles of direction*
agree exactly, in every instance, with the *observed angles of direction*. The comparison of these angles having been made, it was
easy to see the discrepancies; and by shifting the assumed
point, these differences could be reduced to their minimum
value. The point which gave the smallest differences between
the observed and computed angles would be the one towards
which the solar system was progressing. Such was the reasoning
of Argelander, and such the train of investigation on which he
relied for the resolution of this great problem.

Having closed his examinations based on the group of stars
with the most rapid motion, and having found the point in the
heavens which corresponded to their motions, he proceeded to
execute his calculations with reference to his second group. The
stars of this group moved annually an amount greater than half
a second of space, and less than one second. The result was
again reached, and the direction of the solar motion thus derived
agreed, in a remarkable manner, with that obtained from the
first group. A further confirmation was obtained by executing
the calculation founded on the motions of the third and last
group into which he had divided his five hundred stars. The
final result settled, probably for ever, the grand fact that the
sun, with its entire cometary and planetary system, is sweeping
through space towards a point whose place must fall somewhere
within the circumference of a circle whose diameter is about
equal to four times that of the moon.

The reality of the solar motion once determined, astronomers
have not been wanting to verify and extend this wonderful
examination. Argelander's results have been confirmed by the
investigations of M. Otho Struve, the son of the distinguished
director of the Imperial Observatory of Pulkova; and if, on any
fair night, you direct your eye to the constellation of Hercules,
and select from its stars the two marked on the globe with the
Greek-letters π and μ, on the line joining these stars, and at a
distance from π equal to one-quarter of the distance which divides
the stars, will be found the point towards which the sun was
directing his course in the year 1840.

Having obtained the direction of the solar motion, we proceed
to investigate its actual velocity. How swiftly does the sun,
with its retinue of worlds, sweep onward through space? It will
not be possible to present here even an outline of the reasoning

of Struve in the resolution of this intricate question. Two points are involved. The determination of the annual angular motion of the sun, as it would be seen by a spectator situated at a distance equal to that of the stars of the first magnitude. This being determined, the angular motion can readily be converted into linear velocity, in case the mean distance of the stars of the first magnitude can be satisfactorily obtained

After an elaborate investigation, guarded by every care, and open, as it would appear, to no well-founded objections, M. Otho Struve has finally resolved the first of these wonderful questions. It is curious to see how nearly the results agree, which were obtained from data entirely different, and in no way dependent on each other.

M. Otho Struve.

By an examination based on observed right ascensions of the stars, he finds that the space passed over by the sun in its progressive movement through the heavens, seen from the mean distance of stars of the first magnitude, is 321-thousandths of a second of arc. The result obtained from observed declinations gave for the same quantity 357-thousandths of one second of arc. Here is a difference amounting to only 36-thousandths of a second,—a quantity exceedingly small, when we consider the extraordinary difficulty of the investigation.

Let us now convert these numbers into intelligible quantities. In case the sun be supposed to be revolving about some mighty centre, at a distance equal to the mean distance of stars of the first magnitude, the period necessary to accomplish its stupendous revolution will be 3,811,000 years!

Vast as this period appears, we shall see hereafter that we have no right to suppose that the centre about which the solar system is revolving, can be located at a distance nearly so small as the mean distance of the larger stars. But what is the actual velocity? How many miles does this mighty assemblage of flying worlds accomplish in its unknown journey in every year? This is the last question, and even this has not escaped the successful examinations of the human mind. The discovery of the parallax of one or two fixed stars has already been referred to. Within a few months an elaborate work, by Struve, on the Sidereal Heavens, has reached us, containing some remarkable investigations on the mean distances of the stars of the various magnitudes.

Struve, by a most ingenious and powerful train of investigation, obtains a series representing the *relative* mean distances of the stars of all magnitudes, up to the most minute visible in Herschel's twenty-feet reflector. From the sun, as a centre, he sweeps successive concentric spheres, between whose surfaces he conceives the stars of the several magnitudes to be included. The radius of the first sphere reaches to the *nearest* stars of the first magnitude ; that of the second sphere extends to the *farthest* stars of the same magnitude ; and the mean of these two radii will be the mean distance of the stars of the first magnitude. The same is true with reference to the concentric spheres embracing within their surfaces the stars of the various orders of brightness.

Having, from his data, computed a table exhibiting the relative distances of the stars of the different magnitudes, an examination of these figures revealed the singular fact that they constituted a regular geometrical progression ; and having assumed the distance of the stars of the sixth magnitude as the unit, the distance of the stars of the fourth magnitude will be *one half ;* that of those of the second magnitude will be *one quarter,* and so of the even numbers expressing magnitude ; while the distance of the stars of the *fifth* magnitude is obtained by dividing unity by the square root of the number 2, and from this the distances of the odd magnitudes come by dividing constantly by 2. In mathematical language, the distances of the stars of the various magnitudes form a geometrical progression whose ratio is equal to unity divided by the square root of 2.

Having thus obtained the *relative* mean distances of the stars, in case we can find the absolute mean distance of those of any one class, that will reveal to us the absolute mean distances of the stars of every class. For the approximate accomplishment of this last great object, we are again indebted to the astronomers of Russia. As early as 1808, M. Struve, then of Dorpat, attempted the determination of the parallax of a large number of stars, and obtained results so small that, in the state of astronomical science as it then existed, no confidence could be placed in them. The final value of the numerical co-efficient of the aberration of light had not been then absolutely determined. Subsequent investigations by Struve and Peters have fixed this quantity, and the actual determination of the parallax of eight stars recently, has shown that confidence may now be placed in the results obtained by Struve nearly twenty-five years ago.

By combining all the results, M. Peters finds no less than thirty-five stars whose parallaxes have now been determined, either absolute or relative, with a degree of accuracy which warrants their employment in investigating the problem of the mean parallax of stars of the second magnitude. Excluding from this number the stars 61 Cygni, and No. 1830 of the Grombridge catalogue, on account of their great proper motion, there remained thirty-three stars to be employed in the investigation.

From a full and intricate examination of all the data, by a process of reasoning which I will not attempt to explain at this time, M. Peters finds the mean parallax of stars of the *second* magnitude to be equal to 116-thousandths of one second of arc, with a probable error less than a *tenth* part of this quantity. Returning now, with this absolute result, to the table of the relative distances of the fixed stars of different magnitudes, it is easy to fix their absolute distances, as far as confidence can be placed in this first approximation. We find the stars of the first magnitude to be located between the surface of two spheres, whose radii are respectively 986,000 times the radius of the earth's orbit, and 1,246,000 times the same unit. We will express the distance in terms of the velocity of light, as no numbers can convey any intelligible idea. Stars of the first magnitude send us their light in about seventeen years ; those of the second magnitude in about thirty years ; stars of .the third magnitude send their light in about forty-five years ; those of the fourth magnitude in sixty-five years ; those of the fifth in ninety years ; those of the sixth magnitude, the most remote visible to the naked eye, send us their light after a journey through space of one hundred and thirty years ! while the distance of the

lowest order of telescopic stars visible in Herschel's twenty-feet
reflector is such, that their light does not reach the eye for 3541
years after it starts on its tremendous journey!

Let it be remembered that these results are not conjectures.
Though they are first approximations to the truth, they are
reliable to within the tenth part of their value, and are thus far
certain; they raise, in the most astonishing manner, our views
of the immensity of the universe, and of the powers of human
genius which have fathomed these vast and overwhelming
profundities.

Let us now return to the examination of the absolute amount
of progressive motion of our sun and system through space. As
already stated, M. Otho Struve determined its yearly angular
motion, as seen from the more distant of the stars of the first
magnitude. To convert this angular motion into miles, a knowledge must be obtained of the absolute mean distance of the stars
of the first magnitude. This has been accomplished by M. Peters;
and, combining the researches of Argelander, Struve, and Peters,
we are now able to pronounce the following wonderful results:
*The sun, attended by all its planets, satellites, and comets, is sweeping
through space towards the star marked τ in the constellation Hercules,
with a velocity which causes it to pass over a distance equal to
33,350,000 miles in every year!*

And now do you demand how much reliance is to be placed on
this bewildering announcement? I answer, that as to the reality
of the solar motion, there is but one chance out of 400,000 that
astronomers have been deceived. We cannot resist the evidence,
and, startling as the truth appears, we are obliged to yield our
assent, reluctant though it may be, to the logical reasoning by
which this magnificent result has been demonstrated.

But whither is our system tending? If moving onward with
such tremendous velocity, is there not danger that ere long it
may reach the region of the fixed stars, and by sweeping near to
other suns and systems, derange the order of the planetary worlds?
Let us examine this question for one moment, on the hypothesis
that the sun alone is moving among all the stars of heaven, and
that it will hold on in its present direction until it shall reach
the star in Hercules, towards which it is now urging its flight.
This star is of the third magnitude, and, according to our statement already made, the mean distance of its class is such, that
its light does not reach us in a period less than forty-six years.

Executing the calculation, we find that in case the solar system should continue to progress towards that star, it cannot pass the enormous interval, even at 33,550,000 miles per annum, in less than 1,800,000 years !

If the eye of any superior intelligence can behold this amazing scene, how stupendous must be the spectacle presented ! In the centre the sun, blazing with splendour, pursues its majestic career ; around it roll the planets, and about it cluster ten thousand fiery comets. Worlds bright and beautiful hover near the sun ; worlds fiery and chaotic seek this great centre with impetuous velocity, and then dash away into the farthest range of their grand revolution. But the monarch moves on ; and his magnificent cortège, performing his high behests, follow, whithersoever he leads through space !

Here we reach the boundary which divides the known from the unknown. Steadily we have pursued the human mind, as it has moved; on in its grand researches of the universe of God. Time, and space, and number, and distance, have all been set at defiance. No limits have been sufficiently great to circumscribe its movement. For more than 6000 years, onward! has truly been the word. And here I might very well pause, and rest content in the exhibition of the absolute and actual triumphs of human genius ; but as the rays of the rising sun penetrate the darkness of night, and, scattering the gloom, dimly reveal the scenes of earth which are soon to be flooded with splendour, so the light of human knowledge breaks over the boundaries which divide the known from the unknown, and faintly reveals what yet lies far beyond in the dark profound.

Guided by this light, we shall ask your attention to one of the most sublime speculations to which the mind of man has ever risen. I refer to the supposed discovery of the great centre about which it is presumed the myriads of stars composing our mighty Milky Way are all revolving.

M. Maedler, the author of the recent investigations with reference to the *Central Sun*, has long been known to the astronomical world as the successor of M. Struve in the direction of the observatory at Dorpat. His computations of the orbitual movements of the double stars have given to him a deservedly high celebrity, and the great theory which he has propounded is only given to the world after a long and patient examination, extending through seven years.

The extension of the law of gravitation to the fixed stars, now absolutely demonstrated in the revolutions of the binary systems, settles for ever the fact, that in the grand association of stars composing our cluster, or, as we shall hereafter call it, our *astral system*, there must be a *centre of gravity*, as certainly as there is one to the solar system. In the organisation of the solar system we find a central body of vast size, surrounded by small and subordinate satellites. Again, among the planets, we find their magnitude very great, when compared with the moons which circulate around them. Extending this analogy, early astronomers conceived that this principle of a great central preponderating globe would, in all probability, obtain in all the higher orders of physical organisation.

This idea, apparently so well founded, was entirely destroyed by the discovery of the binary stars. Here we find the next higher organisation above our solar system; but instead of finding in the bodies thus united a vast preponderance in magnitude of one over the other, there are many examples in which the two suns thus united by gravitation are in all respects equal. In many others the difference is only slight, yet in all these higher systems there must exist a common centre of gravity.

With the mind cleared, by these views, from all prejudice in favour of the necessary existence of some stupendous central globe, as far exceeding in magnitude the myriads of fixed stars by which it is surrounded as does the sun all the satellites of its system, we are prepared to inquire into the actual existence or non-existence of such a body.

Admitting its invisibility, either in consequence of its distance or non-luminous character, there are yet remaining the means, not only of detecting its existence, but of discovering its position in space. In case such a body exists, the stars located nearest to it will be most completely subjected to its influence, and will show their proximity by the swiftness of their motion. Since it is possible to penetrate space in every direction, in case the stars of any particular region were endowed with a more rapid motion than all others, these would not fail to be discovered. But no such rapid motions have ever been detected, and hence it is now fair to conclude that such motions do not exist, and consequently no vast central globe can ever be found, because there is no evidence that such a body has any locality in space.

The question resolves itself, then, into a research for the

common centre of gravity of all the stars composing our astral system, and the data for such an examination must be found in the direction of the solar motion, and in that of the proper motion of the fixed stars. Difficult as this research undoubtedly is, Maedler's sagacity detected various guides which limited his more minute examinations to a comparatively small portion of the heavens. Since our great astral system has been shown to take the form of a layer or stratum whose thickness is small compared with its extent, we cannot fail to perceive that the centre of gravity of a mass of stars thus arranged must be found somewhere within the limits of the Milky Way, when seen by an eye located not very distant from the centre. But it is seen that our sun does not occupy the absolute centre of this stratum. In case it did, then would the bright circle of the Milky Way divide the heavens into two equal hemispheres. Since there is a manifest difference between the two parts into which the heavens is divided, the smaller portion will be the more distant from us, and in this smaller part we must look for the central point. But, from the soundings of both the Herschels, it is certain that our sun lies nearer the southern half of the Milky Way than the northern. Hence, in our researches for the centre of gravity, we may confine our examinations to the northern half of the smaller of the two parts into which the Milky Way divides the heavens.

One more approximation may be made. If we knew that our sun, in its presumed revolution about this great centre, described a circle, and if we knew the plane of this circle, and the direction in which the sun was now moving, a line drawn in that plane from the sun, and in a direction perpendicular to its line of motion, would pass directly through the centre about which it is revolving, and would point as directly to it. Now the direction of the sun's motion is alone determined; but since the centre of gravity must be found somewhere in a line perpendicular to the direction, we must give to this perpendicular all possible positions in space, which will cause it to cut from the celestial sphere the circumference of a great circle, within which the centre of gravity must be found. These limiting considerations brought the distinguished astronomer to a region of the heavens in and about the constellation Taurus.

Here the examination took a more definite and more strictly scientific form. The proper motion of the stars in this region could be anticipated and known, at least in character and direction. The great centre would probably be located within the limits of some rich cluster. All the stars composing this cluster, as well as those within 20° or 30°, would appear to move in the same direction. Those immediately proximate to the central sun or star would appear to move with the same velocity due to that star, and the entire group would sweep, apparently, through space without parting company.

Having, by such like considerations, narrowed down the limits of research, Maedler commenced his individual examinations. Among other objects subjected to rigid scrutiny, was the brilliant star Aldebaran, in the eye of the Bull. This being the brightest star in this region, and being, moreover, in the midst of a group of smaller stars, seemed, in the outset, to fulfil some of the conditions required of the central sun. But a more rigid examination proved conclusively that this star could not occupy the centre. Its own proper motion far exceeded that of the surrounding stars, and demonstrated its near proximity to our own system, and its mere optical connection with the stars surrounding it.

Thus did this great astronomer move from point to point, from star to star, subjecting each successively to the severest tests, until, finally, a point was found ; a star was discovered, fulfilling, in the most remarkable manner, all the requisitions demanded by the nature of the problem. All are familiar with the beautiful little cluster, called the Pleiades, or seven stars. Clustered around the brilliant star Alcyone, which occupies the optical centre of the group, the telescope shows fourteen conspicuous stars. The proper motions of all these have been determined with great exactitude. These are *all in the same direction*, and are all nearly equal to each other ; and, what is still more important, the mean of their proper motions differs from that of the central star, Alcyone, by only one-thousandth of a second of arc in right ascension, and by two-thousandths of a second in declination. Here, then, is a magnificent group of suns, either actually allied together, and sweeping in company through space, or else they compose a cluster so situated as to be affected by the same apparent motion produced by the sun's progressive motion through the celestial regions.

But an extension of the limits of research around Alcyone exhibits the wonderful truth, that out of one hundred and ten stars within 15° of this centre, there are sixty moving south, or in accordance with the hypothesis that Alcyone is the centre, forty-nine exhibiting no well defined motion, and only one single individual which appears to move contrary to the computed direction !

It is impossible, here, to do justice to the profound and elaborate investigations of the learned author of this great speculation. Assuming Alcyone as the grand centre of the millions of stars composing our astral system, and the direction of the sun's motion, as determined by Argelander and Struve, he investigates the consequent movements of all the stars in every quarter of the heavens. Just where the swiftest motions should be found, there they actually exist, either demonstrating the truth of the theory, or exhibiting the most remarkable and incredible coincidences. We shall not pursue the research. After a profound examination, Maedler reaches the conclusion, *that Alcyone, the principal star in the group of the Pleiades, now occupies the centre of gravity, and is at present the sun about which the universe of stars composing our astral system are all revolving.*

Here, then, we stand on the confines of the unknown. One mighty effort has thus been made to bring beauty and order out of the chaos of motion which has hitherto distinguished the stars of heaven. Once the planets, freed from law, darted through space, or relaxing their speed, actually turned back on their unknown routes. Chaos reigned among these flying globes until the mind, rising by the efforts of its own genius, reached the grand centre of the planetary orbs, and lo ! confusion ceased, and harmony and beauty held their sway among these circling worlds. The same daring human genius which, sweeping across the interplanetary spaces, finally reached the controlling centre of our own great system, has now boldly plunged into the depths of space, has swept across the interstellar spaces, and roaming from star to star, from sun to sun, from system to system, looks out upon the universe of stars, and seeks that point from whence these millions of sweeping suns shall exhibit that grand and magnificent harmony which doubtless reigns throughout the vast empire of Jehovah.

We are too apt to turn away from the first efforts to resolve these mighty problems. How were the doctrines of Newton received ? How much regard was paid to Herschel's grand theory of the solar motion ? And yet how triumphantly have these great theories been established. But do you inquire if there be any possibility of proving or disproving the doctrines of Maedler ? The answer is simple. Should the time ever come when the direction of the solar motion shall be sensibly changed, in consequence of its curvilinear character, then will the plane in which this movement lies be revealed, and then the centre about which the revolution is performed must be made known, at least in direction. Should the line reaching towards this grand centre pass through Alcyone, this, added to all the other evidences, will fix for ever the question of its central position. We know not when this great question may be settled, but judging from the triumphs which have marked the career of human genius hitherto, we do not dare to doubt of the final result.

Admitting the truth of Maedler's theory, we are led to some of the most astonishing results. The known parallax of certain fixed stars gives to us an approximate value of the parallax of Alcyone, and reveals to us the distance of the grand centre. Such is the enormous interval separating the sun from the central star about which it performs its mighty revolution, that the light from Alcyone requires a period of 537 years to traverse the distance ! And if we are to rely on the angular motion of the sun and system, as already determined, at the end of 18,200,000 years, this great luminary, with all its planets, satellites, and comets, will have completed *one* revolution around its grand centre !

Look out to-night on the brilliant constellations which crowd the heavens. Mark the configurations of these stars. Five thousand years ago the Chaldean shepherd gazed on the same bright groups. Two thousand years have rolled away since the Greek philosopher pronounced the eternity of the heavens, and pointed to the ever-during configuration of the stars as proof positive of his assertion. But a time will come when not a constellation now blazing in the bright concave above us shall remain. Slowly, indeed, do these fingers on the dial of heaven mark the progress of time. A thousand years may roll away with scarce a perceptible change, even a million of years may pass without effacing all traces of the groupings which now exist; but that eye which shall behold the universe of the fixed stars when ten millions of years shall have silently rolled away, will search in vain for the constellations which now beautify and adorn our nocturnal heavens. Should God permit, the stars may be there, but no trace of their former relative positions will be found.

Here I must close. The intellectual power of man, as exhibited in his wonderful achievements among the planetary and stellar worlds, has thus far been our single object. I have neither turned to the right hand nor to the left. Commencing with the first mute gaze bestowed upon the heavens, and with the curiosity awakened in that hour of admiration and wonder, we have attempted to follow rapidly the career of the human mind, through the long lapse of six thousand years. What a change has this period wrought !

Go backwards in imagination to the plains of Shinar, and stand beside the shepherd astronomer as he vainly attempts to grasp the mysteries of the waxing and waning moon, and then enter the sacred precincts of yonder temple devoted to the science of the stars. Look over its magnificent machinery ; examine its space-annihilating instruments, and ask the sentinel who now keeps his unbroken vigil, the nature of his investigations.

Moon, and planet, and sun, and system, are left behind. His researches are now within a sphere to whose confines the eagle glance of the Chaldean never reached. Periods, and distances, and masses, and motions, are all familiar to him ; and could the man who gazed and pondered six thousand years ago stand beside the man who now fills his place and listen to his teachings, he would listen with awe, inspired by the revelations of an angel of God. But where does the human mind now stand ? Great as are its achievements, profoundly as it has penetrated the mysteries of creation, what has been done is but an infinitesimal portion of what remains to be done.

But the examinations of the past inspire the highest hopes for the future. The movement is one constantly accelerating and expanding. Look at what has been done during the last three hundred years, and answer me to what point will human genius ascend before the same period shall again roll away ? But in our admiration for that genius which has been able to reveal the mysteries of the universe, let us not forget the homage due to Him who created, and by the might of his power sustains all things. If there be anything which can lead the mind up to the Omnipotent Ruler of the universe, and give to it an approximate knowledge of his incomprehensible attributes, it is to be found in the grandeur and beauty of his works.

If you would know his *glory*, examine the interminable range of suns and systems which crowd the Milky Way. Multiply the

hundred millions of stars which belong to our own "island
universe" by the thousands of these astral systems that exist in
space, within the range of human vision, and then you may form
some idea of the infinitude of his kingdom; for lo! these are but
a part of his ways. Examine the scale on which the universe is
built. Comprehend, if you can, the vast dimensions of our sun.
Stretch outward through his system, from planet to planet, and
circumscribe the whole within the immense circumference of
Neptune's orbit. This is but a single unit out of the myriads of
similar systems. Take the wings of light, and flash with impetuous speed, day and night, and month and year, till youth shall
wear away, and middle age is gone, and the extremest limit of
human life has been attained; count every pulse, and at each,
speed on your way a hundred thousand miles; and when a
hundred years have rolled by, look out, and behold! the thronging
millions of blazing suns are still around you, each separated from
the other by such a distance that in this journey of a century
you have only left half a score behind you.

Would you gather some idea of the *eternity* past of God's
existence? go to the astronomer, and bid him lead you with him
in one of his walks through space; and as he sweeps onward
from object to object, from universe to universe, remember that
the light from those filmy stains on the deep pure blue of heaven,
now falling on your eye, has been traversing space for a million
of years. Would you gather some knowledge of the *omnipotence*
of God? weigh the earth on which we dwell, then count the
millions of its inhabitants that have come and gone for the last
six thousand years. Unite their strength into one arm, and test
its power in an effort to move this earth. It could not stir it a
single foot in a thousand years; and yet under the omnipotent
hand of God, not a minute passes that it does not fly for more
than a thousand miles. But this is a mere atom; the most
insignificant point among his innumerable worlds. At his
bidding, every planet, and satellite, and comet, and the sun himself, fly onward in their appointed courses. His single arm
guides the millions of sweeping suns, and around his throne
circles the great constellation of unnumbered universes.

Would you comprehend the idea of the *omniscience* of God?
remember that the highest pinnacle of knowledge reached by the
whole human race, by the combined efforts of its brightest intellects, has enabled the astronomer to compute approximately the

perturbations of the planetary worlds. He has predicted roughly the return of half a score of comets. But God has computed the mutual perturbations of millions of suns, and planets, and comets, and worlds, without number, through the ages that are passed, and throughout the ages which are yet to come, not approximately, but with perfect and absolute precision. The universe is in motion, system rising above system, cluster above cluster, nebula above nebula, all majestically sweeping around under the providence of God, who alone knows the end from the beginning, and before whose glory and power all intelligent beings, whether in heaven or on earth, should bow with humility and awe.

Would you gain some idea of the *wisdom* of God? look to the admirable adjustments of the magnificent retinue of planets and satellites which sweep around the sun. Every globe has been weighed and poised, every orbit has been measured and bent to its beautiful form. All is changing, but the laws fixed by the wisdom of God, though they permit the rocking to and fro of the system, never introduce disorder, or lead to destruction. All is perfect and harmonious, and the music of the spheres that burn and roll around our sun, is echoed by that of ten millions of moving worlds, that sing and shine around the bright suns that reign above.

If overwhelmed with the grandeur and majesty of the universe of God, we are led to exclaim with the Hebrew poet king, "When I consider thy heavens, the work of thy fingers, the moon and the stars which thou hast ordained, what is man, that thou art mindful of him? and the son of man that thou visitest him?" If fearful that the eye of God may overlook us in the immensity of his kingdom, we have only to call to mind that other passage, " Yet thou hast made him but a little lower than the angels, and hast crowned him with glory and honour. Thou madest him to have dominion over all the works of thy hand; thou hast put all things under his feet." Such are the teachings of the word, and such are the lessons of the works of God.

APPENDIX.

FROM THE "MECHANISM OF THE HEAVENS,"

BY DENISON OLMSTED, LL.D., PROFESSOR OF NATURAL HISTORY AND ASTRONOMY
AT YALE COLLEGE.

THE TELESCOPE.

THE *Telescope,* as its name implies, is an instrument employed for viewing distant objects.* It aids the eye in two ways; first, by enlarging the visual angle under which objects are seen, and, secondly, by collecting and conveying to the eye a much larger amount of the light that emanates from the object, than would enter the naked pupil. A complete knowledge of the telescope cannot be acquired without an acquaintance with the science of optics; but one unacquainted with that science may obtain some idea of the leading principles of this noble instrument. Its main principle is as follows: *By means of the telescope, we first form an image of a distant object,—as the moon for example,—and then magnify that image by a microscope.*

The invention of this noble instrument is generally ascribed to the great philosopher of Florence, Galileo. He had heard that a spectacle-maker of Holland had accidentally hit upon a dis-

* From two Greek words, τηλε, (tele,) *far,* and σκοπεω (skopeo,) *to see.*

covery, by which distant objects might be brought apparently nearer ; and, without further information, he pursued the inquiry, in order to ascertain what forms and combinations of glasses would produce such a result. By a very philosophical process of reasoning, he was led to the discovery of that peculiar form of the telescope which bears his name.

Although the telescopes made by Galileo were no larger than a common glass of the kind now used on board of ships, yet, as they gave new views of the heavenly bodies, revealing the mountains and valleys of the moon, the satellites of Jupiter, and multitudes of stars which are invisible to the naked eye, the discovery was regarded with infinite delight and astonishment.

Reflecting telescopes were first constructed by Sir Isaac Newton, although the use of a concave reflector, instead of an object-glass, to form the image, had been previously suggested by Gregory, an eminent Scottish astronomer, whose name is still employed to designate the Gregorian telescope. The first telescope made by Newton was only six inches long, and its reflector was little more than an inch in diameter. Notwithstanding its small dimensions, it performed so well, as to encourage further efforts ; and this illustrious philosopher afterwards constructed much larger instruments, one of which, made with his own hands, was presented to the Royal Society of London, and is now carefully preserved in the library of the Society.

Newton was induced to undertake the construction of reflecting telescopes, from the belief that refracting telescopes were necessarily limited to a very small size, with only moderate illuminating powers ; whereas the dimensions and powers of the former admitted of being indefinitely increased. Considerable *magnifying* powers might, indeed, be obtained from refractors, by making them very long ; but the *brightness* with which telescopic objects are seen, depends greatly on the dimensions of the beam of light which is collected by the object-glass, or by the mirror, and conveyed to the eye ; and therefore small object-glasses cannot have a very high illuminating power. The experiments of Newton on colours led him to believe, that it would be impossible to employ large lenses in the construction of telescopes, since such glasses would give to the images they formed the colours of the rainbow. But later opticians have found means of correcting these imperfections, so that we are now able to use object-glasses a foot or more in diameter, which give very clear and bright images. Such instruments are called *achromatic* telescopes,—a name implying the absence of prismatic or rainbow colours in the image. It is, however, far more difficult to construct large achromatic than large reflecting telescopes. Very large pieces of glass can seldom be found that are sufficiently pure for the purpose; since every inequality in the glass, such as waves, tears, threads, and the like, spoils it for optical purposes, as they distort the light, and produce confused images.

The achromatic telescope (that is, the refracting telescope, having such an object-glass as to give a colourless image) was invented by Dollond, a distinguished London artist, about the year 1757. He had in his possession a quantity of glass of a remarkably fine quality, which enabled him to carry his invention at once to a high degree of perfection. It has ever since been a matter of the greatest difficulty, with the manufacturers of telescopes, to find pieces of glass, of a suitable quality for object-glasses, more than two or three inches in diameter. Hence, large achromatic telescopes are very expensive, being valued in proportion to the *cubes* of their diameters; that is, if a telescope whose aperture (as the breadth of the object-glass is technically called) is two inches, cost twenty-four pounds, one whose aperture is eight inches would cost one thousand five hundred and twenty pounds.

Since it is so much easier to make large reflecting than large refracting telescopes, it may be asked, why the latter are ever attempted, and why reflectors are not exclusively employed? I answer, that the achromatic telescope, when large and well constructed, is a more perfect and more durable instrument than the reflecting telescope. Much more of the light that falls on the mirror is absorbed than is lost in passing through the object-glass of a refractor; and hence the larger achromatic telescopes afford a stronger light than the reflecting, unless the latter are made of an enormous and unwieldy size. Moreover, the mirror is very liable to tarnish, and will never retain its full lustre for many years together; and it is no easy matter to restore the lustre when once impaired.

The three most celebrated telescopes hitherto made, are Herschel's *forty-feet reflector*, the *great Dorpat refractor*, and the still more remarkable telescope recently completed by Lord Rosse. Herschel was a Hanoverian by birth, but settled in England in the younger part of his life. As early as 1774, he began to make telescopes for his own use ; and, during his life, he made more than four hundred, of various sizes and powers. Under the patronage of George III., he completed, in 1789, his great telescope, having a tube of iron forty feet long, and a speculum forty-nine and a half inches, or more than four feet in diameter. Let us endeavour to form a just conception of this gigantic instrument, which we can do only by dwelling on its dimensions, and comparing them with those of other objects with which we are familiar, as the length or height of a house, and the breadth of a hogshead. The reflector alone weighed nearly a ton. So large and ponderous an instrument must require a vast deal of machinery to work it, and to keep it steady; and accordingly the frame-work surrounding it was formed of heavy timbers, and resembled the frame of a large building. When one of the largest of the fixed stars, as Sirius, is entering the field of this telescope, its approach is announced by a bright dawn, like that which precedes the rising sun ; and when the star itself enters the field, the light is insupport-able to the naked eye. The planets are expanded into brilliant luminaries, like the moon; and innumerable multitudes of stars are scattered like glittering dust over the celestial vault.

The great Dorpat telescope is of more recent construction. It was made by Fraunhofer, a German optician, of the greatest eminence, at Munich, in Bavaria, and takes its name from being attached to the observatory at Dorpat, in Russia. It is of much smaller dimensions than the great telescope of Herschel. Its object-glass is nine and a half inches in diameter, and its length fourteen feet. Although the price of this instrument was nearly one thousand two hundred pounds sterling, yet it is said that this sum barely covered the actual expenses. It weighs five thousand pounds, and yet is turned with the finger. In facility of management it has greatly the advantage of Herschel's tele-scope. Moreover, the sky of England is so frequently unfavour-able for astronomical observation, that *one hundred* good hours (or those in which the higher powers can be used) are all that can be obtained in a whole year. On this account, as well as from the difficulty of shifting the position of the instrument, Herschel estimated that it would take about six hundred years to obtain with it even a momentary glimpse of every part of the heavens. This remark shows that such great telescopes are unsuited to the common purposes of astronomical observation.

Indeed, most of Herschel's discoveries were made with his small telescopes; and although, for certain rare purposes, powers were applied which magnified seven thousand times, yet, in most of his observations, powers magnifying only two or three hundred times were sufficient. The highest power of the Dorpat telescope is only seven hundred, and yet the director of this instrument, Professor Struve, is of opinion that it is nearly or quite equal in quality, all things considered, to Herschel's forty-feet reflector.*

* The largest and most remarkable telescope ever constructed, is that which science owes to the enterprise, public spirit, and enlightened love of knowledge which distinguishes the Earl of Rosse. This noble instrument, which is erected at Parsonstown, in Ireland, is a *reflecting* telescope. "The great speculum is 6 feet diameter 5½ inches thick at the edges, and 5 inches at the centre, and its weight is about 3 *tons*. Its composition is copper and tin; 126 parts of copper to 57½ of tin. The price of the copper alone was reckoned at about £100. By grinding and polishing, its thickness was reduced about one eighth of an inch. Its focal distance is about 54 feet. The casting of the speculum took place in April, 1842, which, with all the multifarious operations connected with it, was accomplished without any accident, and with a degree of success beyond expectation. The speculum has a reflecting surface of 4071 square inches, while that of Sir W. Herschel's 40 feet telescope had only 1811 square inches on its polished surface; so that the quantity of light reflected from this speculum is considerably more than double that of Herschel's largest reflector, and it is chiefly on the quantity of light either transmitted or reflected, that the power of telescopes to penetrate into space depends. The process of grinding this speculum was conducted under water, and the moving power employed was a steam-engine of three horse power. The substance made use of to wear down the surface was emery and water, a constant supply of these was kept between the grinder and the speculum. It required six weeks to grind it to a fair surface.

" The tube of this telescope is 56 feet long, including the speculum box, and is made of deal, 1 inch thick, hooped with iron. On the inside, at intervals of 8 feet, there are rings of iron, 3 inches in depth, and 1 inch broad, for the purpose of strengthening the sides. The diameter of the tube is 7 feet. It is fixed to mason-work in the ground, to a large universal hinge; which allows it to turn in all directions. At 12 feet distance on each side, a wall is built, 72 feet long, 48 high on the outer side, and 56 on the inner, the walls being 24 feet distance from each other, and lying exactly in the meridional line. When directed to the south, the tube may be lowered till it become almost horizontal; but when pointed to the north, it only falls till it is parallel to the earth's axis, pointing then to the pole of the heavens. Its lateral movements take place only from wall to wall: and this commands a view for half-an-hour on each side of the meridian; that is, the whole of its motion from east to west is limited to 15 degrees. It is ultimately intended to connect with the tube-end galleries, machinery which shall give an automaton movement, so that the telescope shall be used as an equatorial instrument. The tube and speculum, including the bed on which the speculum rests, weigh about 15 tons.

" The telescope rests on a universal joint, placed on masonry about 6 feet below the ground, and it is elevated and depressed by a chain and windlass; and though it weighs 15 tons, the instrument is raised by two men with great facility. Of course it is counterpoised in every direction. The observer, when at work, stands in one of four galleries, the three highest of which are drawn out from the western wall, while the fourth, or lowest, has for its base an elevating platform, along the horizontal surface of which a gallery slides from wall to wall, by machinery within the observer's reach, but which a child may work. When the telescope is directed to an object near the zenith, the observer stands at an elevation at least 50 feet above the ground. The

Still further improvements, however, have since greatly increased the power of this remarkable astronomical instrument, and chiefly by the combination of the Le Mairean, with the Newtonian principle. Sir James South illustrates the value which such an improvement would give, in the following remarks on Lord Rosse's Telescope :—" Thus the difficulty of constructing a Newtonian telescope, of dimensions never before contemplated, is completely overcome ; but to render the part on which I am about to enter more generally intelligible, let me say, that the Newtonian telescope is composed of a large concave speculum, of a small flat speculum, and of an eye-glass. The large concave speculum lies in the closed end of the tube, at right angles to the tube's axis. The small flat speculum is placed near the open end of the tube in its centre, but at half right angles with the tube ; whilst the eye-glass (a hole for the purpose being pierced in the tube's side) is fixed opposite the centre of the flat speculum. The rays from the object to which the telescope is directed fall on the large concave speculum, are reflected from it into a point called the focus, in which the image of the object is formed ; this image falls on the flat speculum, and is reflected from it to the eye-glass, by which it becomes magnified, and enters the observer's eye. But only a part of the light which falls on the large concave speculum is reflected on the small speculum ; and again, only a part of that which falls from the large speculum on the small one, is reflected from the latter to the eye-glass. Newton, to avoid the loss of light by the second reflection, proposed the substitution of a glass prism for the small flat speculum ; but from some difficulties which have attended its use, it has (perhaps too hastily) been laid aside.

" In 1728, Le Maire presented to the Académie des Sciences the plan of a reflecting telescope, in which the use of the small flat speculum was suppressed ; for by giving the large concave speculum a little inclination, he threw the image formed in its focus near to one side of the tube, where an eye-glass magnifying it, the observer viewed it, his back at the time being turned

engraving represents only the upper part of the tube of the telescope, at which the observer stands when making his observations; and as the telescope is at present of the Newtonian construction, he looks in at the upper part of the *side* of the great tube; but it is intended to throw aside the plane speculum, and to adopt the *front view*; when the observer will look directly down the tube on the image formed by the great speculum."

towards the object in the heavens; thus the light lost in the Newtonian telescope by the second reflection was saved.

"No one, however, seems to have availed himself of this form of construction till 1776, when Herschel, with a ten-feet reflector, tried it, and rejected it. In 1784, he again tried it with a twenty-feet reflector, but again abandoned it. In 1786 he adopted it, eulogised it very much, among other advantages, for giving almost double the light of the Newtonian construction, and called it 'the front view.' Subsequently to this period all his twenty-feet telescopes, as well as his forty-feet telescope, were constructed as Le Maireans.

"The excess of light, however, is very much overrrated by Sir W. Herschel; for experiments since made indicate that the diameter of the Newtonian telescope must be increased about one-fifth only to obtain equal light with one of the Le Mairean construction.

"That we might have a practical proof of the advantages of the light of the Le Mairean construction, the three feet Newtonian of twenty-seven feet focus, which stands in the demesne by the side of the Leviathan, was temporarily fitted up as a Le Mairean. Stars of the first magnitude were seen, not well defined, as in the Newtonian form of the instrument; but the superiority of the Le Mairean, where a large quantity of light was required, was most decided. The small pole-star was as bright as a star of the fourth magnitude when seen in an eight feet achromatic of three and three quarter inches aperture. The dumb-bell nebulæ, or 27 of Messier, was resolved into clusters of stars in a manner never before seen with it. The annular nebulæ of Lyra, brilliant beyond what it had ever yet appeared, was surrounded by stars too bright to escape immediate notice, although neither the dumb-bell nebulæ nor the annular nebulæ had more than fifteen degrees of altitude when I placed the telescope on them.

"On the 15th of March, when the moon was seven days and a half old, I never saw her unillumined disc so beautifully, nor her mountains so temptingly measureable. On my first looking into the telescope, a star of about the seventh magnitude was some minutes of a degree distant from the moon's dark limb. Seeing that its occultation by the moon was inevitable, as it was the first occultation which had been observed by that telescope, I was anxious that it should be observed by its noble maker; and very much do I regret that, through kindness towards me, he would not accede to my wish ; for the star, instead of disappearing the moment the moon's edge came in contact with it, apparently glided on the moon's dark face, as if it had been seen through a transparent moon, or as if the star were between me and the moon. It remained on the moon's disk nearly two seconds of time, and then instantly disappeared, at 10h. 9m. 50·72s. sidereal time. I have seen this apparent projection of a star on the moon's face several times, but from the great brilliancy of the star, this was the most beautiful I ever saw." Sir James South thus proceeds to compare the great telescopes :—

"The only telescopes, in point of size, comparable with Lord Rosse's three feet and six feet, are Sir William Herschel's twenty feet and forty feet Le Mairean's. The twenty feet had a speculum of 18·8 inches diameter, and the forty feet one of four feet.

"The Le Mairean of 18·8 inches diameter, in point of light is equal to a Newtonian of 22½ inches diameter.

"The Le Mairean of four feet diameter is equal to a Newtonian of $57\frac{4}{10}$ inches.

"The Le Mairean of three feet is equal to a Newtonian of 43 inches.

"And the Le Mairean of six feet is equal to a Newtonian of 86 inches.

"By substituting, then, the Le Mairean form for the Newtonian, the present three feet Newtonian will be made as effective as if it were 43 inches diameter, and the six feet as if it were 86 inches in diameter ; or the quantity of light in each telescope, after the alteration, will be, to its present light, as seven to five nearly, or almost half as much again as it now has.

"Seeing, then, that the change from the Newtonian to the Le Mairean construction will be attended with such an accession of light, Lord Rosse, having determined geometrically the form of the curve requisite to produce with it a definition of objects equal to that which each of the telescopes at present gives, is devising mechanical means for producing it."

It is not generally understood in what way greatness of size in a telescope increases its powers ; and it conveys but an imperfect idea of the excellence of a telescope to tell how much it magnifies. In the same instrument, an increase of magnifying power is always attended with a diminution of the light and of the field of view. Hence, the lower powers generally afford the most agreeable views, because they give the clearest light and take in the largest space. The several circumstances which influence the qualities of a telescope are illuminating power, magnifying power, distinctness, and field of view. Large mirrors and large object-glasses are superior to smaller ones, because they collect a larger beam of light, and transmit it to the eye. Stars which are invisible to the naked eye are rendered visible by the telescope, because this instrument collects and conveys to the eye a large beam of the few rays which emanate from the stars ; whereas a beam of these rays of only the diameter of the pupil of the eye, would afford too little light for distinct vision. In this particular, large telescopes have great advantages over small ones. The great mirror of Herschel's forty feet reflector collect and conveys to the eye a beam more than four feet in diameter. The Dorpat telescope also transmits to the eye a beam nine and a half inches in diameter. This seems small, in comparison with the reflector ; but much less of the light is lost on passing through the glass than is absorbed by the mirror, and the mirror is very liable to be clouded or tarnished; so that there is not so great a difference in the two instruments, in regard to illuminating power, as might be supposed from the difference of size.

Distinctness of view is all-important to the performance of an instrument. The object may be sufficiently bright, yet, if the image is distorted, or ill-defined, the illumination is of little consequence. This property depends mainly on the skill with which all the imperfections of figure and colour in the glass or mirror are corrected, and can exist in perfection only when the image is rendered completely achromatic, and when all the rays that proceed from each point in the object are collected into corresponding points of the image, unaccompanied by any other rays. Distinctness is very much affected by the *steadiness* of the instrument. Every one knows how indistinct a page becomes when a book is passed rapidly backwards and forwards before the eyes, and how difficult it is to read in a carriage in rapid motion on a rough road.

Field of view is another important consideration. The finest instruments exhibit the moon, for example, not only bright and distinct, in all its parts, but they take in the whole disc at once ; whereas the inferior instruments, when the higher powers especially are applied, permit us to see only a small part of the moon at once.

Some states of the weather, even when the sky is clear, are far more favourable for astronomical observations than others. After sudden changes of temperature in the atmosphere, the medium is usually very unsteady. If the sun shines out warm after a cloudy season, the ground first becomes heated, and the air that is nearest to it is expanded, and rises, while the colder air descends, and thus ascending and descending currents of air, mingling together, create a confused and wavy medium. The same cause operates when a current of hot air rises from a chimney; and hence the state of the atmosphere in cities and large towns is very unfavourable to the astronomer on this account, as well as on account of the smoky condition in which it is usually found. After a long season of dry weather also the air becomes smoky, and unfit for observation. Indeed, foggy, misty, or smoky air, is so prevalent in some countries, that only a very few times in the whole year can be found, which are entirely suited to observation, especially with the higher powers; for we must recollect that these inequalities and imperfections are magnified by telescopes as well as the objects themselves. Thus, as I have already mentioned, not more than one hundred good hours in a year could be obtained for observation with Herschel's great telescope. By *good* hours, Herschel means that the sky must be very clear, the moon absent, no twilight, no haziness, no violent wind, and no sudden change of temperature. As a general fact, the warmer climates enjoy a much finer sky for the astronomer than the colder, having many more clear evenings, a short twilight, and less change of temperature. The watery vapour of the atmosphere also is more perfectly dissolved in hot than in cold air, and the more water air contains, provided it is in a state of perfect solution, the clearer it is.

A *certain preparation of the observer himself* is also requisite for the nicest observations with the telescope. He must be free from all agitation, and the eye must not recently have been exposed to a strong light, which contracts the pupil of the eye. Indeed, for delicate observations, the observer should remain for some time beforehand in a dark room, to let the pupil of the eye dilate. By this means, it will be enabled to admit a larger number of the rays of light. In ascending the stairs of an observatory, visitors frequently get out of breath, and having perhaps recently emerged from a strongly lighted apartment, the eye is not in a favourable state for observation. Under these disadvantages, they take a hasty look into the telescope, and it is no wonder that disappointment usually follows.

Want of steadiness is a great difficulty attending the use of the highest magnifiers; for the motions of the instrument are magnified as well as the object. Hence, in the structure of observatories, the greatest pains is requisite, to avoid all tremor, and to give to the instruments all possible steadiness; and the same care is to be exercised by observers. In the more refined observations, only one or two persons ought to be near the instrument.

In general, *low powers* afford better views of the heavenly bodies than very high magnifiers. It may be thought absurd to recommend the use of low powers, in respect to large instruments especially, since it is commonly supposed that the advantage of large instruments is, that they will bear high magnifying powers. But this is not their only, nor even their principal advantage. A good light and large field are qualities, for most purposes, more important than great magnifying power; and it must be borne in mind that as we increase the magnifying power in a given instrument, we diminish both the illumination and the field of view. Still, different objects require different magnifying powers; and a telescope is usually furnished with several varieties of powers, one of which is best fitted for viewing the moon, another for Jupiter, and a still higher power for Saturn. Comets require only the lowest magnifiers; for here, our object is to command as much light, and as large a field as possible, while it avails little to increase the dimensions of the object. On the other hand, for certain double stars (stars which appear single to the naked eye, but double to the telescope), we require very high magnifiers, in order to separate these minute objects so far from each other that the interval can be distinctly seen. Whenever we exhibit celestial objects to inexperienced observers, it is usual to precede the view with good *drawings* of the objects, accompanied by an explanation of what each appearance, exhibited in the telescope, indicates. The novice is told that mountains and valleys can be seen in the moon by the aid of the telescope; but, on looking, he sees a confused mass of light and shade, and nothing which looks to him like either mountains or valleys. Had his attention been previously directed to a plain drawing of the moon, and each particular appearance interpreted to him, he would then have looked through the telescope with intelligence and satisfaction.

OBSERVATORIES.

An observatory is a structure fitted up expressly for astronomical observations, and furnished with suitable instruments for that purpose.

The two most celebrated observatories hitherto built are those of Tycho Brahe, and of Greenwich. The observatory of Tycho Brahe was constructed at the expense of the King of Denmark, in a style of royal magnificence, and cost no less than two hundred thousand crowns. It was situated on the island of Huenna, at the entrance of the Baltic, and was called Uraniburg, or the palace of the skies.

Before giving an account of Tycho's observatory, it will be useful to relate a few particulars respecting this great astronomer himself.

Tycho Brahe was of Swedish descent, and of noble family ; but having received his education at the University of Copenhagen, and spent a large part of his life in Denmark, he is usually considered a Dane, and quoted as a Danish astronomer. He was born in the year 1546. When he was about fourteen years old, there happened a great eclipse of the sun, which awakened in him a high interest, especially when he saw how accurately all the circumstances of it answered to the prediction with which he had been before made acquainted. He was immediately seized with an irresistible passion to acquire a knowledge of the science which could so successfully lift the veil of futurity. His friends had destined him for the profession of law, and, from the superior talents of which he gave early promise, added to the advantage of powerful family connexions, they had marked out for him a distinguished career in public life. They therefore endeavoured to discourage him from pursuing a path which they deemed so much less glorious, and vainly sought, by various means, to extinguish the zeal for astronomy which was kindled in his youthful bosom. Despising all the attractions of a court, he contracted an alliance with a peasant girl, and in the peaceful retirement of domestic life desired no happier lot than to peruse the grand volume which the nocturnal heavens displayed to his enthusiastic imagination. He soon established his fame as one of the greatest astronomers of the age, and monarchs did homage to his genius. The King of Denmark became his munificent patron, and James I., King of England, when he went to Denmark to complete his marriage with a Danish princess, passed eight days with Tycho in his observatory, and, at his departure, addressed to the astronomer a Latin ode, accompanied with a magnificent present. He gave him also his royal license to print his works in England, and added to it the following complimentary letter :—" Nor am I acquainted with these things on the relation of others, or from a mere perusal of your works, but I have seen them with my own eyes, and heard them with my own ears, in your own residence at Uraniburg, during the various learned and agreeable conversations which I there held with you, which even now affect my mind to such a degree, that it is difficult to decide whether I recollect them with greater pleasure or admiration." Admiring disciples also crowded to this sanctuary of the sciences, to acquire a knowledge of the heavens.

The observatory consisted of a main building, which was square, each side being sixty feet, and of large wings in the form of round towers. The whole was executed in a style of great magnificence ; and Tycho, who was a nobleman by descent, gratified his taste for splendour and ornament by giving to every part of the structure an air of the most finished elegance. Nor were the instruments with which it was furnished less magnificent than the buildings. They were vastly larger than had before been employed in the survey of the heavens, and many of them were adorned with costly ornaments. One of Tycho's large and splendid instruments was his astronomical quadrant, on one side of which was figured a representation of the astronomer and his assistants in the midst of their instruments, and intently engaged in making and recording observations. The description of this instrument, furnished by his contemporaries, conveys to us a striking idea of the magnificence of his arrangements, and of the extent of his operations.

Here Tycho sat in state, clad in the robes of nobility, and supported throughout his establishment the etiquette due to his rank. His observations were more numerous than all that had ever been made before, and they were carried to a degree of accuracy that is astonishing, when we consider that they were made without the use of the telescope, which was not yet invented.

Tycho carried on his observations at Uraniburg for about twenty years, during which time he accumulated an immense store of accurate and valuable facts, which afforded the ground-work of the discovery of the great laws of the solar system established by Kepler.

But the high marks of distinction which Tycho enjoyed, not only from his own sovereign, but also from foreign potentates, provoked the envy of the courtiers of his royal patron. They did not, indeed, venture to make their attacks upon him while his generous patron was living; but the king was no sooner dead, and succeeded by a young monarch, who did not feel the same interest in protecting and encouraging this great ornament of the kingdom, than his envious foes carried into execution their long-meditated plot for his ruin. They represented to the young king that the treasury was exhausted, and that it was necessary to retrench a number of pensions, which had been granted for useless purposes, and in particular that of Tycho, which, they maintained, ought to be conferred upon some person capable of rendering greater services to the state. By these means they succeeded in depriving him of his support, and he was compelled to withdraw to the hospitable mansion of a friend in Germany. Here he became known to the Emperor, who invited him to Prague, where, with an ample stipend, he resumed his labours. But, though surrounded with affectionate friends and admiring disciples, he was still an exile in a foreign land. Although his country had been base in its ingratitude, it was yet the land which he loved; the scene of his earliest affection; the theatre of his scientific glory. These feelings continually preyed upon his mind, and his unsettled spirit was ever hovering among his native mountains. In this condition he was attacked by a disease of the most painful kind; and though its agonising paroxysms had lengthened intermissions, yet he saw that death was approaching. He implored his pupils to persevere in their scientific labours; he conversed with Kepler on some of the profoundest points of astronomy; and with these secular occupations he mingled frequent acts of piety and devotion. In this happy condition he expired, without pain, at the age of fifty-five.*

The observatory at Greenwich was not built until a hundred years after that of Tycho Brahe, namely, in 1676. The great interests of the British nation, which are involved in navigation, constituted the ruling motive with the government to lend their aid in erecting and maintaining this observatory.

The site of the observatory at Greenwich is on a commanding eminence facing the river Thames, five miles east of the central parts of London. Being part of a royal park, the neighbouring grounds are in no danger of being occupied by buildings, so as to obstruct the prospect. It is also in full view of the shipping on the Thames ; and, according to a standing regulation, at the instant of one o'clock, every day, a huge ball is dropped from the top of a staff on the observatory, as a signal to the commanders of vessels for regulating their chronometers.

The buildings comprise a series of rooms, of sufficient number and extent to accommodate the different instruments, the inmates of the establishment, and the library ; and on the top is a celebrated camera obscura, exhibiting a most distinct and perfect picture of the grand and unrivalled scenery which this eminence commands.

This establishment, by the accuracy and extent of its observations, has contributed more than all other institutions to perfect the science of astronomy.

To preside over and direct this great institution, a man of the highest eminence in the science is appointed by the government, with the title of *Astronomer Royal.* He is paid an ample salary, with the understanding that he is to devote himself exclusively to the business of the observatory. The astronomers royal of the Greenwich observatory, from the time of its first establishment in 1676, to the present time, have constituted a series of the proudest names of which British science can boast.

Six assistants, besides inferior labourers, are constantly in attendance ; and the business of making and recording observations is conducted with the utmost system and order.

The great objects to be attained in the construction of an observatory are, a commanding and unobstructed view of the heavens ; freedom from causes that affect the transparency and uniform state of the atmosphere, such as fires, smoke, or marshy

* Brewster's Life of Newton.

grounds ; mechanical facilities for the management of instruments, and especially every precaution that is necessary to secure perfect steadiness. This last consideration is one of the greatest importance, particularly in the use of very large magnifiers ; for we must recollect, that any motion in the instrument is magnified by the full power of the glass, and gives a proportional unsteadiness to the object. A situation is, therefore, selected as remote as possible from public roads (for even the passing of carriages would give a tremulous motion to the ground, which would be sensible in large instruments), and structures of solid masonry are commenced deep enough in the ground to be unaffected by frost, and built up to the height required, without any connexion with the other parts of the building. Many observatories are furnished with a moveable dome for a roof, capable of revolving on rollers, so that instruments penetrating through the roof may be easily brought to bear upon any point, at or near the zenith.

It will not, perhaps, be desired that I should go into a minute description of all the various instruments that are used in a well-constructed observatory. Nor is this necessary, since a very large proportion of all astronomical observations are taken on the meridian, by means of the transit instrument and clock. When a body, in its diurnal revolution, comes to the meridian, it is at its highest point above the horizon, and is then least affected by refraction and parallax. This, then, is the most favourable position for taking observations upon it. Moreover, it is peculiarly easy to take observations on a body when in this situation. Hence the transit instrument and clock are the most important members of an astronomical observatory, and some account of these instruments becomes indispensable.

The *transit instrument* is a telescope which is fixed permanently in the meridian, and moves only in that plane. It can, therefore, be so directed, as to observe the passage of a star across the meridian at any altitude. The accompanying graduated circle enables the observer to set the instrument at any required altitude, corresponding to the known altitude at which the body to be observed crosses the meridian. Or it may be used to measure the altitude of a body, or its zenith distance, at the time of its meridian passage. Near the circle may be seen a spirit-level, which serves to show when the axis is exactly on a level with the horizon. The frame-work is made of solid metal (usually brass), everything being arranged with reference to keeping the instrument perfectly steady. It stands on screws, which not only afford a steady support, but are useful for adjusting the instrument to a perfect level. The transit instrument is sometimes fixed immoveably to a solid foundation, as a pillar of stone, which is built up from a depth in the ground below the reach of frost. When enclosed in a building, as in an observatory, the stone pillar is carried up separate from the walls and floors of the building, so as to be entirely free from the agitations to which they are liable.

The use of the transit instrument is to show the precise instant when a heavenly body is on the meridian, or to measure the time it occupies in crossing the meridian. The *astronomical clock* is the constant companion of the transit instrument. This clock is so regulated as to keep exact pace with the stars, and, of course, with the revolution of the earth on its axis : that is, it is regulated to *sidereal* time. It measures the progress of a star, indicating an hour for every fifteen degrees, and twenty-four hours for the whole period of the revolution of the star. Sidereal time commences when the vernal equinox is on the meridian, just as solar time commences when the sun is on the meridian. Hence the hour by the sidereal clock has no correspondence with the hour of the day, but simply indicates how long it is since the equinoctial point crossed the meridian. For example, the clock of an observatory points to three hours and twenty minutes ; this may be in the morning, at noon, or any other time of the day,—for it merely shows that it is three hours and twenty minutes since the equinox was on the meridian. Hence, when a star is on the meridian, the clock itself shows its right ascension, which, it will be recollected, is the angular distance measured on the equinoctial, from the point of intersection of the ecliptic and equinoctial, called the vernal equinox, reckoning fifteen degrees for every hour, and a proportional number of degrees and minutes for a less period. I have before remarked, that a very large portion of all astronomical observations are taken when the bodies are on the meridian, by means of the transit instrument and clock.

Having now described these instruments, I will next explain the manner of using them for different observations. Anything becomes a measure of time which divides duration equally. The equinoctial, therefore, is peculiarly adapted to this purpose, since, in the daily revolutions of the heavens, equal portions of the equinoctial pass under the meridian in equal times. The only difficulty is, to ascertain the amount of these portions for given intervals. Now, the clock shows us exactly this amount; for, when regulated to sidereal time (as it easily may be), the hour-hand keeps exact pace with the equator, revolving once on the dial-plate of the clock while the equator turns once by the revolution of the earth. The same is true, also, of all the small circles of diurnal revolution; they all turn exactly at the same rate as the equinoctial, and a star situated anywhere between the equator and the pole will move in its diurnal circle along with the clock, in the same manner as though it were in the equinoctial. Hence, if we note the interval of time between the passage of any two stars, as shown by the clock, we have a measure of the number of degrees by which they are distant from each other in right ascension. Hence we see how easy it is to take arcs of right ascension: the transit instrument shows when a body is on the meridian; the clock indicates how long it is since the vernal equinox passed it, which is the right ascension itself; or it tells the difference of right ascension between any two bodies, simply by indicating the difference in time between their periods of passing the meridian. Again, it is easy to take the *declination* of a body when on the meridian. By declination, the reader will recollect, is meant the distance of a heavenly body from the equinoctial; the same, indeed, as latitude on the earth. When a star is passing the meridian, if, on the instant of crossing the meridian wire of the telescope, we take its distance from the north pole (which may readily be done, because the position of the pole is always known, being equal to the latitude of the place), and subtract this distance from ninety degrees, the remainder will be the distance from the equator, which is the declination. It will be asked, why we take this indirect method of finding the declination? Why we do not rather take the distance of the star from the equinoctial at once? I answer, that it is easy to point an instrument to the north pole, and to ascertain its exact position, and, of course, to measure any distance from it on the meridian; while, as there is nothing to mark the exact situation of the equinoctial, it is not so easy to take direct measurements from it. When we have thus determined the situation of a heavenly body, with respect to two great circles at right angles with each other, as in the present case, the distance of a body from the equator and from the equinoctial colure, or that meridian which passes through the vernal equinox, we know its relative position in the

heavens ; and when we have thus determined the relative positions of all the stars, we may lay them down on a map or a
globe, exactly as we do places on the earth, by means of their
latitude and longitude.

The foregoing is only a *specimen* of the various uses of the
transit instrument, in finding the relative places of the heavenly
bodies. Another use of this excellent instrument is, to regulate
our clocks and watches. By an observation with the transit
instrument, we find when the sun's centre is on the meridian.
This is the exact time of *apparent* noon. But watches and clocks
usually keep *mean* time, and therefore, in order to set our time-
piece by the transit instrument, we must apply to the apparent
time of noon the equation of time, as is explained in the next
chapter.

A *noon-mark* may easily be made by the aid of the transit
instrument. A window sill is frequently selected as a suitable
place for the mark, advantage being taken of the shadow projected upon it by the perpendicular casing of the window. Let
an assistant stand, with a rule laid on the line of shadow, and
with a knife ready to mark, the instant when the observer at the
transit instrument announces that the centre of the sun is on the
meridian. By a concerted signal, as the stroke of a bell, the
inhabitants of a town may all fix a noon-mark from the same
observation. If the signal be given on one of the days when
apparent time and mean time become equal to each other, as on
the twenty-fourth of December, no equation of time is required.

As a noon-mark is convenient for regulating time-pieces, I will
point out a method of making one, which may be practised without the aid of the telescope. Upon a smooth, level plane, freely
exposed to the sun, with a pair of compasses describe a circle.
In the centre, where the leg of the compasses stood, erect a
perpendicular wire of such a length, that the termination of its
shadow shall fall upon the circumference of the circle at some
hour before noon, as about ten o'clock. Make a small dot at the
point where the end of the shadow falls upon the circle, and do
the same where it falls upon it again in the afternoon. Take a
point half-way between these two points, and from it draw a line
to the centre, and it will be a true meridian line. The direction
of this line would be the same, whether it were made in the
summer or in the winter ; but it is expedient to draw it about
the fifteenth of June, for then the shadow alters its length most
rapidly, and the moment of its crossing the wire will be more

definite, than in the winter. At this time o year, also, the sun
and clock agree, or are together, as is more fully explained in the
next chapter ; whereas, at other times of the year, the time of
noon, as indicated by a common clock, would not agree with that
indicated by the sun. If the upper end of the wire is flattened
and a small hole is made in it, through which the sun may shine
the instant when this bright spot falls upon the circle will be
better defined than the termination of the shadow.

Another important instrument of the observatory is the *mural
circle*. It is a graduated circle, usually of very large size, fixed
permanently in the plane of the meridian, and attached firmly to
a perpendicular wall ; and on its centre is a telescope, which
revolves along with it, and is easily brought to bear on any
object in any point in the meridian. It is made of large size,
sometimes twenty feet in diameter, in order that very small
angles may be measured on its limb ; for it is obvious that a
small angle, as one second, will be a larger space on the limb of
an instrument, in proportion as the instrument itself is larger.
The vertical circle usually connected with the transit instrument,
may indeed be employed for the same purposes as the mural
circle, namely, to measure arcs of the meridian, as meridian
altitudes, zenith distances, north polar distances, and declinations ; but as that circle must necessarily be small, and therefore
incapable of measuring very minute angles, the mural circle is
particularly useful in measuring these important arcs. It is very
difficult to keep so large an instrument perfectly steady ; and
therefore it is attached to a massive wall of solid masonry, and
is hence called a *mural* circle, from the Latin word *murus*, which
signifies a wall.

Every expedient is employed to give the instrument firmness
of parts and steadiness of position. The circle is of solid metal,
usually of brass, and it is strengthened by numerous radii, which
keep it from warping or bending ; and these are made in the
form of hollow cones, because that is the figure which unites in
the highest degree lightness and strength. On the rim of the
instrument is a microscope. This is attached to a micrometer,—
a declicate piece of apparatus, used for reading the minute sub-
divisions of angles ; for, after dividing the limb of the instrument
as minutely as possible, it will then be necessary to magnify
those divisions with the microscope, and subdivide each of these
parts with the micrometer. Thus, if we have a mural circle
twenty feet in diameter, and of course nearly sixty-three feet in

circumference, since there are twenty-one thousand and six hundred minutes in the whole circle, we shall find, by calculation, that one minute would occupy, on the limb of such an instrument, only about one thirtieth of an inch, and a second, only one eighteen hundredth of an inch. We could not, therefore, hope to carry the actual divisions to a greater degree of minuteness than minutes ; but each of these spaces may again be subdivided into seconds by the micrometer.

From these statements, the reader will acquire some faint idea of the extreme difficulty of making perfect astronomical instruments, especially where they are intended to measure such minute angles as one second. Indeed, the art of constructing astronomical instruments is one which requires such refined mechanical genius, —so superior a mind to devise, and so delicate a hand to execute, —that the most celebrated instrument-makers take rank with the most distinguished astronomers ; supplying, as they do, the means by which only the latter are enabled to make these great discoveries. Astronomers have sometimes made their own telescopes ; but they have seldom, if ever, possessed the refined manual skill which is requisite for graduating delicate instruments.

Lord Orrery.

TIME AND THE CALENDAR.

THE reader who has hitherto been conversant only with the many fine and sentimental things which the poets have sung respecting Old Time, will, perhaps find some difficulty in bringing down his mind to the calmer consideration of what time really is, and according to what different standards it is measured for different purposes. He will not, however, I think, find the subject even in our matter-of-fact and unpoetical way of treating it, altogether uninteresting. What, then, is time ? *Time is a measured portion of indefinite duration.* It consists of equal portions cut off from eternity, as a line on the surface of the earth is separated from the contiguous portions which constitute a great circle of the sphere, by applying to it a two-foot scale ; or as a few yards of cloth are measured off from a web of unknown or indefinite extent.

Any thing, or any event which takes place at equal intervals, may become a measure of time. Thus, the pulsations of the wrist, the flowing of a given quantity of sand from one vessel to another, as in an hour-glass, the beating of a pendulum, and the revolution of a star, have been severally employed as measures of time. But the great standard of time is the period of the revolution of the earth on its axis, which, by the most exact observations, is found to be always the same. I have anticipated this subject in some degree, in giving an account of the transit instrument and clock, but I propose here, to enter into it more at large.

The time of the earth's revolution on its axis, as already explained, is called a sidereal day, and is determined by the revolution of a star in the heavens. This interval is divided into twenty-four *sidereal* hours. Observations taken on numerous stars, in different ages of the world, show that they all perform their diurnal revolution in the same time, and that their motion, during any part of the revolution, is always uniform. Here, then, we have an exact measure of time, probably more exact than any thing which can be devised by art. *Solar time* is reckoned by the apparent revolution of the sun from the meridian round to the meridian again. Were the sun stationary in the heavens, like a fixed star, the time of its apparent revolution would be equal to the revolution of the earth on its axis, and the solar and the sidereal days would be equal. But, since the sun passes from west to east, through three hundred and sixty degrees, in three hundred and sixty-five and one-fourth days, it moves eastward nearly one degree a day. While, therefore, the earth is turning round on its axis, the sun is moving in the same direction, so that, when we have come round under the same celestial meridian from which we started, we do not find the sun there, but he has moved eastward nearly a degree, and the earth must perform so much more than one complete revolution, before we come under the sun again. Now, since we move, in the diurnal revolution, fifteen degrees in sixty minutes, we must pass over one degree in four minutes. It takes, therefore, four minutes for us to overtake the sun, after we have made one complete revolution. Hence the solar day is about four minutes longer than the sidereal ; and if we were to reckon the sidereal day twenty-four hours, we should reckon the solar day twenty-four hours four minutes. To suit the ordinary purposes of society, however, it is found more convenient to reckon the solar days twenty-four hours, and throw the fraction into the sidereal day. Then,

$$24\text{h. } 4\text{m. } : 24\text{h. } : : 24\text{h. } : 23\text{h. } 56\text{m. } 4\text{s.}$$

That is, when we reduce twenty-four hours and four minutes to twenty-four hours, the same proportion will require that we reduce the sidereal day from twenty-four hours to twenty-three hours fifty-six minutes four seconds ; or, in other words, a sidereal day is such a part of a solar day. The solar days, however, do not always differ from the sidereal by precisely the same fraction, since they are not constantly of the same length. Time, as measured by the sun, is called *apparent time,* and a clock so regulated as always to keep exactly with the sun, is said to keep apparent time. *Mean time* is reckoned by the *average* length of all the solar days throughout the year. This is the period which constitutes the *civil* day of twenty-four hours, beginning when the sun is on the lower meridian, namely, at twelve o'clock at night, and counted by twelve hours from the lower to the upper meridian, and from the upper to the lower. The *astronomical* day is the apparent solar day counted through the whole twenty-four hours, (instead of by periods of twelve hours each, as in the civil day,) and begins at noon. Thus, when it is the tenth of June, at nine o'clock, A.M., according to civil time, the tenth of June has not yet commenced by astronomical time, nor will it, until noon; consequently, it is then June ninth, twenty-first hour of astronomical time. Astronomers, since so many of their observations are taken on the meridian, are always supposed to look towards the south. Geographers, having formerly been conversant only with the northern hemisphere, are always understood to be looking towards the north. Hence, left and right, when applied to the astronomer, mean east and west, respectively; but to the geographer the right is east, and the left, west.

Clocks are usually regulated so as to indicate mean solar time; yet, as this is an artificial period not marked off, like the sidereal day, by any natural event, it is necessary to know how much is to be added to, or subtracted from, the apparent solar time, in order to give the corresponding mean time. The interval, by which apparent time differs from mean time, is called the *equation of time*. If one clock is so constructed as to keep exactly with the sun, going faster or slower, according as the lengths of the solar days vary, and another clock is regulated to mean time, then the difference of the two clocks, at any period, would be the equation of time for that moment. If the apparent clock be *faster* than the mean, then the equation of time must be subtracted; but if the apparent clock be slower than the mean, then the equation of time must be added, to give the mean time. The two clocks would differ most about the third of November, when the apparent time is sixteen and one-fourth minutes greater than the mean. But since apparent time is sometimes greater and sometimes less than mean time, the two must obviously be sometimes equal to each other. This is the case four times a year, namely, April fifteenth, June fifteenth, September first, and December twenty-fourth.

Astronomical clocks are made of the best workmanship, with

every advantage that can promote their regularity. Although they are brought to an astonishing degree of accuracy, yet they are not as regular in their movements as the stars are, and their accuracy requires to be frequently tested. The transit instrument itself, when once accurately placed in the meridian, affords the means of testing the correctness of the clock, since one revolution of a star, from the meridian to the meridian again, ought to correspond exactly to twenty-four hours by the clock, and to continue the same, from day to day; and the right ascensions of various stars, as they cross the meridian, ought to be such by the clock, as they are given in the tables, where they are stated according to the accurate determinations of astronomers. Or, by taking the difference of any two stars, on successive days, it will be seen whether the going of the clock is uniform for that part of the day; and by taking the right ascensions of different pairs of stars, we may learn the rate of the clock at various parts of the day. We thus learn, not only whether the clock accurately measures the length of the sidereal day, but also whether it goes uniformly from hour to hour.

Although astronomical clocks have been brought to a great degree of perfection, so as hardly to vary a second for many months, yet none are absolutely perfect, and most are so far from it, as to require to be corrected by means of the transit instrument, every few days. Indeed, for the nicest observations, it is usual not to attempt to bring the clock to a state of absolute correctness, but, after bringing it as near to such a state as can conveniently be done, to ascertain how much it gains or loses in a day; that is, to ascertain the *rate* of its going, and to make allowance accordingly.

Having considered the manner in which the smaller divisions of time are measured, let us now take a hasty glance at the larger periods which compose the calendar.

As a *day* is the period of the revolution of the earth on its axis, so a *year* is the period of the revolution of the earth around the sun. This time, which constitutes the *astronomical year*, has been ascertained with great exactness, and found to be three hundred and sixty-five days five hours forty-eight minutes and fifty-one seconds. The most ancient nations determined the number of days in the year by means of the *stylus*, a perpendicular rod which casts its shadow on a smooth plane bearing a meridian line. The time when the shadow was shortest, would indicate the day of the summer solstice; and the number of days which elapsed, until the shadow returned to the same length again, would show the number of days in the year. This was found to be three hundred and sixty-five whole days, and accordingly this period was adopted for the civil year. Such a difference, however, between the civil and astronomical years, at length threw all dates into confusion. For if, at first, the summer solstice happened on the twenty-first of June, at the end of four years the sun would not have reached the solstice until the twenty-second of June; that is, it would have been behind its time. At the end of the next four years, the solstice would fall on the twenty-third; and in process of time it would fall successively on every day of the year. The same would be true of any other fixed date.

Julius Cæsar, who was distinguished alike for the variety and extent of his knowledge, and his skill in arms, first attempted to make the calendar conform to the motions of the sun:

> " Amidst the hurry of tumultuous war,
> The stars, the gods, the heavens, were still his care."

Aided by Sosigenes, an Egyptian astronomer, he made the first correction of the calendar, by introducing an additional day every fourth year, making February to consist of twenty-nine instead of twenty-eight days, and of course the whole year to consist of three hundred and sixty-six days. This fourth year was denominated *Bissextile*, because the sixth day before the Kalends of March was reckoned twice. It is also called Leap Year.

The Julian year was introduced into all the civilised nations that submitted to the Roman power, and continued in general use until the year 1582. But the true correction was not six hours, but five hours forty-nine minutes; hence the addition was too great by eleven minutes. This small fraction would amount in one hundred years to three-fourths of a day, and in one thousand years to more than seven days. From the year 325 to the year 1582, it had, in fact, amounted to more than ten days; for it was known that, in 325, the vernal equinox fell on the twenty-first of March, whereas, in 1582, it fell on the eleventh. It was ordered by the Council of Nice, a celebrated ecclesiastical council, held in the year 325, that Easter should be celebrated upon the first Sunday after the first full moon next following

the vernal equinox ; and as certain other festivals of the Romish
Church were appointed at particular seasons of the year, confusion
would result from such a want of constancy between any fixed
date and a particular season of the year. Suppose, for example,
a festival, accompanied by numerous religious ceremonies, was
decreed by the Church to be held at the time when the sun
crossed the equator in the spring (an event hailed with great
joy, as the harbinger of the return of the summer), and that, in
the year 325, March twenty-first was designated as the time for
holding the festival, since, at that period, it was on the twenty-
first of March when the sun reached the equinox ; the next year
the sun would reach the equinox a little sooner than the twenty-
first of March, only eleven minutes, indeed, but still amounting
in twelve hundred years to ten days ; that is, in 1582, the sun
reached the equinox on the eleventh of March. If, therefore,
they should continue to observe the twenty-first as a religious
festival in honour of this event, they would commit the absurdity
of celebrating it ten days after it had passed by. Pope
Gregory XIII., who was then at the head of the Roman see,
was a man of science, and undertook to reform the calendar, so
that fixed dates would always correspond to the same seasons of
the year. He first decreed that the year should be brought
forward ten days, by reckoning the fifth of October the
fifteenth ; and, in order to prevent the calendar from falling into
confusion afterwards, he prescribed the following rule :—*Every
year whose number is not divisible by four, without a remainder,
consists of three hundred and sixty-five days ; every year which
is so divisible, but is not divisible by one hundred, of three
hundred and sixty-six ; every year divisible by one hundred,
but not by four hundred, again, of three hundred and sixty-five ;
and every year divisible by four hundred, of three hundred and
sixty-six.*

Thus the year 1850, not being divisible by four, contains three
hundred and sixty-five days, while 1852 and 1856 are leap years.
Yet, to make every fourth year consist of three hundred and
sixty-six days would increase it too much, by about three-
fourths of a day in a century ; therefore every hundredth year
has only three hundred and sixty-five days. Thus 1800, although
divisible by four, was not a leap year, but a common year. But
we have allowed a *whole* day in a hundred years, whereas we
ought to have allowed only *three-fourths* of a day. Hence, in four

hundred years, we should allow a day too much, and therefore we let the four hundreth remain a leap year. This rule involves an error of less than a day in four thousand two hundred and thirty-seven years.

The Pope, who in that age assumed authority over all secular princes, issued his decree to the reigning sovereigns of Christendom, commanding the observance of the calendar as reformed by him. The decree met with great opposition among the Protestant states, as they recognised in it a new exercise of ecclesiastical tyranny; and some of them, when they received it, made it expressly understood, that their acquiescence should not be construed as a submission to the Papal authority.

In 1752, the Gregorian year, or *New Style*, was established in Great Britain by act of Parliament, and the dates of all deeds, and other legal papers, were to be made according to it. As above a century had then passed since the first introduction of the new style, eleven days were suppressed, the third of September being called the fourteenth. By the same act, the beginning of the year was changed from March twenty-fifth to January first. A few persons born previously to 1752 have survived to our day, and we occasionally see inscriptions on tombstones of those whose time of birth is recorded in old style. In order to make this correspond to our present mode of reckoning, we must add eleven days to the date. Thus the same event would be June twelfth of old style, or June twenty-third of new style; and if an event occurred between January first and March twenty-fifth, the date of the year would be advanced one, since February 1, 1740, old style, would be February 1, 1741, new style. Thus General Washington was born February 11, 1731, old style, or February 22, 1732, new style. If we inquire how any present event may be made to correspond in date to the old style, we must subtract twelve days, and put the year back one, if the event lies between January first and March twenty-fifth. Thus June tenth, new style, corresponds to May twenty-ninth, old style; and March 20, 1840, to March 8, 1839. France, being a a Roman Catholic country, adopted the new style soon after it was decreed by the Pope; but Protestant countries, as we have seen, were much slower in adopting it; and Russia, and the Greek Church generally, still adhere to the old style. In order, therefore, to make the Russian dates correspond to ours, we must add to them twelve days.

It may seem very remarkable that so much pains should have been bestowed upon this subject; but without a correct and uniform standard of time, the dates of deeds, commissions, and all legal papers; of fasts and festivals, appointed by ecclesiastical authority; the returns of seasons, and the records of history,—must all fall into inextricable confusion. To change the observance of certain religious feasts, which have been long fixed to particular days, is looked upon as an impious innovation; and though the times of the events, upon which these ceremonies depend, are utterly unknown, it is still insisted upon by certain classes in England, that the Glastonbury thorn blooms on Christmas-day.

Although the ancient Grecian calendar was extremely defective, yet the common people were entirely averse to its reformation. Their superstitious adherence to these errors was satirised by Aristophanes, in his Comedy of the Clouds. An actor, who had just come from Athens, recounts that he met with Diana, or the moon, and found her extremely incensed, that they did not regulate her course better. She complained that the order of Nature was changed, and every thing turned topsyturvy. The gods no longer knew what belonged to them: but, after paying their visits on certain feast-days, and expecting to meet with good cheer, as usual, they were under the disagreeable necessity of returning back to the skies without their suppers!

Among the Greeks, and other ancient nations, the length of the year was generally regulated by the course of the moon. This planet, on account of the different appearances which she exhibits at her full, change, and quarters, was considered by them as best adapted of any of the celestial bodies for this purpose. As one lunation, or revolution of the moon around the earth, was found to be completed in about twenty-nine and one-half days, twelve of these periods being supposed equal to one revolution of the sun, their months were made to consist of twenty-nine and thirty days alternately, and their year of three hundred and fifty-four days. But this disagreed with the annual revolution of the sun, which must evidently govern the seasons of the year, more than eleven days. The irregularities, which such a mode of reckoning would occasion, must have been too obvious not to have been observed. For, supposing it to have been settled, at any particular time, that the beginning of the year should be in the Spring; in about sixteen years afterwards the beginning would have been in Autumn; and in thirty-three or thirty-four years it would have gone backwards through all the seasons to Spring again. This defect they attempted to rectify, by introducing a number of days, at certain times, into the calendar, as occasion required, and putting the beginning of the year forwards, in order to make it agree with the course of the sun. But as these additions, or *intercalations*, as they were called, were generally left to be regulated by the priests, who, from motives of interest or superstition, frequently omitted them, the year was made long or short at pleasure.

The *week* is another division of time, of the highest antiquity, which, in almost all countries has been made to consist of seven days,—a period supposed by some to have been traditionally derived from the creation of the world; while others imagine it to have been regulated by the phases of the moon. The names, Saturday, Sunday, and Monday, are obviously derived from Saturn, the Sun, and the Moon; while Tuesday, Wednesday, Thursday, and Friday, are the days of Tuisco, Woden, Thor, and Friga, which are Saxon names for deities corresponding to Mars, Mercury, Jupiter, and Venus.*

The common year begins and ends on the same day o the week; but leap year ends one day later than it began. Fifty-two weeks contain three hundred and sixty-four days; if, therefore, the year begins on Tuesday, for example, we should complete fifty-two weeks on Monday, leaving one day (Tuesday) to complete the year, and the following year would begin on Wednesday. Hence, any day of the month is one day later in the week than the corresponding day of the preceding year. Thus, if the sixteenth of November, 1838, falls on Friday, the sixteenth of November, 1837, fell on Thursday, and will fall, in 1839, on Saturday. But if leap year begins on Sunday, it ends on Monday, and the following year begins on Tuesday; while any given day of the month is two days later in the week than the corresponding date of the preceding year.

* Bonnycastle's Astronomy.

COPERNICUS AND GALILEO.

—•—

COPERNICUS, a native of Thorn, in Prussia, was born in 1473.
Though destined for the profession of medicine, from his earliest
years he displayed a great fondness and genius for mathematical
studies, and pursued them with distinguished success in the
University of Cracow. At the age of twenty-five years, he
resorted to Italy, for the purpose of studying astronomy, where
he resided a number of years. Thus prepared, he returned to
his native country, and, having acquired an ecclesiastical living
that was adequate to his support in his frugal mode of life, he
established himself at Frauenberg, a small town near the mouth
of the Vistula, where he spent nearly forty years in observing
the heavens, and meditating on the celestial motions. He occupied the upper part of a humble farm-house, through the roof of
which he could find access to an unobstructed sky, and there he
carried on his observations. His instruments, however, were
few and imperfect, and it does not appear that he added any
thing to the art of practical astronomy. This was reserved for
Tycho Brahe, who came half a century after him. Nor did
Copernicus enrich the science with any important discoveries.
It was not so much his genius or taste to search for new bodies,
or new phenomena among the stars, as it was to explain the
reasons of the most obvious and well-known appearances and
motions of the heavenly bodies. With this view, he gave his
mind to long-continued and profound meditation.

Copernicus tells us that he was first led to think that the
apparent motions of the heavenly bodies, in their diurnal revo-
tion, were owing to the real motion of the earth in the opposite
direction, from observing instances of the same kind among

terrestrial objects ; as when the shore seems to the mariner to recede, as he rapidly sails from it ; and as the trees and other objects seem to glide by us, when, on riding swiftly past them, we lose the consciousness of our own motion. He was also smitten with the *simplicity* prevalent in all the works and operations of nature, which is more and more conspicuous the more they are understood ; and he hence concluded that the planets do not move in the complicated paths which most preceding astronomers assigned to them. I shall explain, hereafter, the details of his system. I need only at present remind the reader that the hypothesis which he espoused and defended, (being substantially the same as that proposed by Pythagoras, five hundred years before the Christian era,) supposes, first, that the apparent movements of the sun by day, and of the moon and stars by night, from east to west, result from the actual revolution of the earth on its own axis from west to east ; and, secondly, that the earth and all the planets revolve about the sun in circular orbits. This hypothesis, when he first assumed it, was with him, as it had been with Pythagoras, little more than mere conjecture. The arguments by which its truth was to be finally established were not yet developed, and could not be, without the aid of the telescope, which was not yet invented. Upon this hypothesis, however, he set out to explain all the phenomena of the visible heavens,—as the diurnal revolutions of the sun, moon, and stars, the slow progress of the planets through the signs of the zodiac, and the numerous irregularities to which the planetary motions are subject. These last are apparently so capricious,—being for some time forward, then stationary, then backward, then stationary again, and finally direct, a second time, to the order of the signs, and constantly varying in the velocity of their movements,—that nothing but long-continued and severe meditation could have solved all these appearances, in conformity with the idea that each planet is pursuing its simple way all the while in a circle around the sun. Although, therefore, Pythagoras fathomed the profound doctrine that the sun is the centre around which the earth and all the planets revolve, yet we have no evidence that he ever solved the irregular motions of the planets in conformity with his hypothesis, although the explanation of the diurnal revolution of the heavens, by that hypothesis, involved no difficulty. Ignorant as Copernicus was of the principle of gravitation, and of most of the laws of motion, he could go but little way in following out the consequences of his own hypothesis; and all that can be claimed for him is, that he solved, by means of it, most of the common phenomena of the celestial motions. He indeed got upon the road to truth, and advanced some way in its sure path; but he was able to adduce but few independent proofs, to show that it was the truth. It was only quite at the close of his life that he published his system to the world, and that only at the urgent request of his friends; anticipating, perhaps, the opposition of a bigoted priesthood, whose fury was afterwards poured upon the head of Galileo, for maintaining the same doctrines.

Although, therefore, the system of Copernicus afforded an explanation of the celestial motions, far more simple and rational than the previous systems which made the earth the centre of these motions, yet this fact alone was not sufficient to compel the assent of astronomers; for the greater part of the same phenomena could be explained with tolerable consistency on either hypothesis. With the old doctrine astronomers were already familiar, a circumstance which made it seem easier; while the new doctrines would seem more difficult, from their being imperfectly understood. Accordingly, for nearly a century after the publication of the system of Copernicus, he gained few disciples. Tycho Brahe rejected it, and proposed one of his own, of which I shall hereafter give some account; and it would probably have fallen quite into oblivion, had not the observations of Galileo, with his newly-invented telescope, brought to light innumerable proofs of its truth, far more cogent than any which Copernicus himself had been able to devise.

Galileo no sooner had completed his telescope, and directed it to the heavens, than a world of wonders suddenly burst upon his enraptured sight. Pointing it to the moon, he was presented with a sight of her mottled disc, and of her mountains and valleys. The sun exhibited his spots; Venus, her phases; and Jupiter, his expanded orb, and his retinue of moons. These last he named, in honour of his patron, Cosmo de' Medici, *Medicean stars*. So great was the honour deemed of associating one's name with the stars, that express application was made to Galileo, by the court of France, to award this distinction to the reigning monarch, Henry IV., with plain intimations, that by so doing he would render himself and his family rich and powerful for ever.

Galileo published the result of his discoveries in a paper, denominated "*Nuncius Sidereus*," the "Messenger of the Stars." In that ignorant and marvellous age, this publication produced a wonderful excitement. "Many doubted, many positively refused to believe, so novel an announcement; all were struck with the greatest astonishment, according to their respective opinions, either at the new view of the universe thus offered to them, or at the high audacity of Galileo, in inventing such fables." Even Kepler, the great German astronomer, to whom I shall refer more particularly by and by, wrote to Galileo, and desired him to supply him with arguments, by which he might answer the objections to these pretended discoveries with which he was continually assailed. Galileo answered him as follows : "In the first place, I return you my thanks that you first, and almost alone, before the question had been sifted, (such is your candour, and the loftiness of your mind,) put faith in my assertions. You tell me you have some telescopes, but not sufficiently good to magnify distant objects with clearness, and that you anxiously expect a sight of mine, which magnifies images more than a thousand times. It is mine no longer, for the Grand Duke of Tuscany has asked it of me, and intends to lay it up in his museum, among his most rare and precious curiosities, in eternal remembrance of the invention.

"You ask, my dear Kepler, for other testimonies. I produce, for one, the Grand Duke, who, after observing the Medicean planets several times with me at Pisa, during the last months, made me a present at parting of more than a thousand florins, and has now invited me to attach myself to him, with the annual salary of one thousand florins, and with the title of 'Philosopher and Principal Mathematician to His Highness;' without the duties of any office to perform, but with the most complete leisure. I produce, for another witness, myself, who, although already endowed in this college with the noble salary of one thousand florins, such as no professor of mathematics ever before received, and which I might securely enjoy during my life, even if these planets should deceive me and should disappear, yet quit this situation, and take me where want and disgrace will be my punishment, should I prove to have been mistaken."

The learned professors in the universities, who, in those days, were unaccustomed to employ their senses in inquiring into the phenomena of nature, but satisfied themselves with the authority of Aristotle, on all subjects, were among the most incredulous with respect to the discoveries of Galileo. " Oh, my dear Kepler," says Galileo, " how I wish that we could have one hearty laugh together. Here, at Padua, is the principal professor of philosophy, whom I have repeatedly and urgently requested to look at the moon and planets through my glass, which he pertinaciously refuses to do. Why are you not here ? What shouts of laughter we should have at this glorious folly, and to hear the Professor of Philosophy at Pisa labouring before the Grand Duke, with logical arguments, as if with magical incantations, to charm the new planets out of the sky."

The following argument by Sizzi, a contemporary astronomer of some note, to prove that there can be only seven planets, is a specimen of the logic with which Galileo was assailed. " There are seven windows given to animals in the domicile of the head, through which the air is admitted to the tabernacle of the body, to enlighten, to warm, and to nourish it ; which windows are the principal parts of the microcosm, or little world,—two nostrils, two eyes, two ears, and one mouth. So in the heavens, as in a macrocosm, or great world, there are two favourable stars, Jupiter and Venus ; two unpropitious, Mars and Saturn ; two luminaries, the Sun and Moon ; and Mercury alone, undecided and indifferent. From which, and from many other phenomena of nature, such as the seven metals, &c., which it were tedious to enumerate, we gather that the number of planets is necessarily seven. Moreover the satellites are invisible to the naked eye, and therefore can exercise no influence over the earth, and therefore would be useless, and therefore do not exist. Besides, as well the Jews and other ancient nations, as modern Europeans, have adopted the division of the week into seven days, and have named them from the seven planets. Now, if we increase the number of planets, this whole system falls to the ground."

When, at length, the astronomers of the schools found it useless to deny the fact that Jupiter is attended by smaller bodies, which revolve around him, they shifted their ground of warfare, and asserted that Galileo had not told the whole truth ; that there were not merely *four* satellites, but a still greater number ; one said five ; another nine ; and another, twelve ; but, in a little time, Jupiter moved forward in his orbit, and left all behind him, save the four Medicean stars.

It had been objected to the Copernican system, that were Venus a body which revolved around the sun in an orbit interior to that of the earth, she would undergo changes similar to those of the moon. As no such changes could be detected by the naked eye, no satisfactory answer could be given to this objection ; but the telescope set all right, by showing, in fact, the phases of Venus. The same instrument disclosed also, in the system of Jupiter and his moons, a miniature exhibition of the solar system itself. It showed the actual existence of the motion of a number of bodies around one central orb, exactly similar to that which was predicated of the sun and planets. Every one, therefore, of these new and interesting discoveries, helped to confirm the truth of the system of Copernicus.

But a fearful cloud was now rising over Galileo, which spread itself, and grew darker every hour. The Church of Rome had taken alarm at the new doctrines respecting the earth's motions, as contrary to the declarations of the Bible, and a formidable difficulty presented itself, namely, how to publish and defend these doctrines, without invoking the terrible punishments inflicted by the Inquisition on heretics. No work could be printed without license from the court of Rome ; and any opinions supposed to be held and much more known to be taught by any one, which, by an ignorant and superstitious priesthood, could be interpreted as contrary to Scripture, would expose the offender to the severest punishments, even to imprisonment, scourging, and death. We, who live in an age so distinguished for freedom of thought and opinion, can form but a very inadequate conception of the bondage in which the minds of men were held by the chains of the Inquisition. It was necessary, therefore, for Galileo to proceed with the greatest caution in promulgating truths which his own discoveries had confirmed. He did not, like the Christian martyrs, proclaim the truth in the face of persecutions and tortures ; but while he sought to give currency to the Copernican doctrines, he laboured at the same time, by cunning artifices, to blind the ecclesiastics to his real designs, and thus to escape the effects of their hostility.

Before Galileo published his doctrines in form, he had expressed himself so freely, as to have excited much alarm among the ecclesiastics. One of them preached publicly against him, taking for his text the passage, " Ye men of Galilee, why stand ye here gazing up into heaven ? " He therefore thought it prudent to

s

resort to Rome, and confront his enemies face to face. A contemporary describes his appearance there in the following terms,
in a letter addressed to a Romish Cardinal : "Your Eminence
would be delighted with Galileo if you heard him holding forth,
as he often does, in the midst of fifteen or twenty, all violently
attacking him, sometimes in one house, sometimes in another.
But he is armed after such fashion, that he laughs all of them to
scorn ; and even if the novelty of his opinions prevents entire
persuasion, at least he convicts of emptiness most of the arguments with which his adversaries endeavour to overwhelm
him."

In 1616, Galileo, as he himself states, had a most gracious
audience of the Pope, Paul V., which lasted for nearly an
hour, at the end of which his Holiness assured him, that the
Congregation were no longer in a humour to listen lightly to
calumnies against him, and that so long as he occupied the Papal
chair, Galileo might think himself out of all danger. Nevertheless, he was not allowed to return home, without receiving formal
notice not to teach the opinions of Copernicus, "that the sun is
in the centre of the system, and that the earth moves about it,"
from that time forward, in any manner.

Galileo had a most sarcastic vein, and often rallied his persecutors with the keenest irony. This he exhibited, some time
after quitting Rome, in an epistle which he sent to the Arch-
duke Leopold, accompanying his "Theory of the Tides." "This
theory," says he, "occurred to me when in Rome, whilst the
theologians were debating on the prohibition of Copernicus's
book, and of the opinion maintained in it of the motion of the
earth, which I at that time believed ; until it pleased those
gentlemen to suspend the book, and to declare the opinion false
and repugnant to the Holy Scriptures. Now, as I know how
well it becomes me to obey and believe the decisions of my
superiors, which proceed out of more profound knowledge than
the weakness of my intellect can attain to, this theory which I
send you, which is founded on the motion of the earth, I now
look upon as a fiction and a dream, and beg your Highness to
receive it as such. But, as poets often learn to prize the crea-
tions of their fancy, so, in like manner, do I set some value on
this absurdity of mine. It is true, that when I sketched this
little work, I did hope that Copernicus would not, after eighty
years, be convicted of error ; and I had intended to develope and

amplify it further; but a voice from heaven suddenly awakened me, and at once annihilated all my confused and entangled fancies."

It is difficult, however, sometimes to decide whether the language of Galileo is ironical, or whether he uses it with subtlety, with the hope of evading the anathemas of the Inquisition. Thus he ends one of his writings with the following passage: "In conclusion, since the motion attributed to the earth, which I, as a pious and Catholic person, consider most false, and not to exist, accommodates itself so well to explain so many and such different phenomena, I shall not feel sure that, false as it is, it may not just as deludingly correspond with the phenomena of comets."

In the year 1624, soon after the accession of Urban VIII. to the pontifical chair, Galileo went to Rome again, to offer his congratulations to the new Pope, as well as to propitiate his favour. He seems to have been received with unexpected cordiality; and on his departure the Pope commended him to the good graces of Ferdinand, Grand Duke of Tuscany, in the following terms: "We find in him not only literary distinction, but the love of piety, and he is strong in those qualities by which Pontifical good-will is easily obtained. And now, when he has been brought to this city to congratulate Us on Our elevation, We have lovingly embraced him; nor can We suffer him to return to the country whither your liberality recals him, without an ample provision of Pontifical love. And that you may know how dear he is to Us, we have willed to give him this honourable testimonial of virtue and piety. And We further signify, that every benefit which you shall confer upon him will conduce to Our gratification."

In the year 1630, Galileo finished a great work, on which he had been long engaged, entitled, "The Dialogue on the Ptolemaic and Copernican systems." From the notion which prevailed, that he still countenanced the Copernican doctrine of the earth's motion, which had been condemned as heretical, it was some time before he could obtain permission from the Inquisitors (whose license was necessary to every book) to publish it. This he was able to do, only by employing again that duplicity or artifice which would throw dust in the eyes of the vain and superstitious priesthood. In 1632, the work appeared under the following title: "A Dialogue, by Galileo Galilei, Extraordinary

Mathematician of the University of Pisa, and Principal Philosopher and Mathematician of the Most Serene Grand Duke of Tuscany; in which, in a conversation of four days, are discussed the two principal Systems of the World, the Ptolemaic and Copernican, indeterminately proposing the Philosophical Arguments as well on one side as on the other." The subtle disguise which he wore may be seen from the following "Introduction," addressed "To the discreet Reader."

" Some years ago, a salutary edict was promulgated at Rome, which, in order to obviate the perilous scandals of the present age, enjoined an opportune silence on the Pythagorean opinion of the earth's motion. Some were not wanting, who rashly asserted that this decree originated not in a judicious examination, but in ill-informed passion; and complaints were heard, that counsellors totally inexperienced in astronomical observations ought not, by hasty prohibitions, to clip the wings of speculative minds. My zeal could not keep silence when I heard these rash lamentations, and I thought it proper, as being fully informed with regard to that most prudent determination, to appear publicly on the theatre of the world, as a witness of the actual truth. I happened at that time to be in Rome: I was admitted to the audiences, and enjoyed the approbation, of the most eminent prelates of that court; nor did the publication of that decree occur without my receiving some prior intimation of it. Wherefore it is my intention, in this present work, to show to foreign nations, that as much as is known of this matter in Italy, and particularly in Rome, as ultramontane diligence can ever have formed any notion of, and collecting together all my own speculations on the Copernican system, to give them to understand that the knowledge of all these preceded the Roman censures; and that from this country proceed not only dogmas for the salvation of the soul, but also ingenious discoveries for the gratification of the understanding. With this object, I have taken up in the 'Dialogue' the Copernican side of the question, treating it as a pure mathematical hypothesis; and endeavouring, in every artificial manner, to represent it as having the advantage, not over the opinion of the stability of the earth absolutely, but according to the manner in which that opinion is defended by some, who indeed profess to be Aristotelians, but retain only the name, and are contented, without improvement, to worship shadows, not philosophising with their own reason, but only from the recollection of the four principles imperfectly understood."

Although the Pope himself, as well as the Inquisitors, had examined Galileo's manuscript, and, not having the sagacity to detect the real motives of the author, had consented to its publication, yet, when the book was out, the enemies of Galileo found means to alarm the court of Rome, and Galileo was summoned to appear before the Inquisition. The philosopher was then seventy years old, and very infirm, and it was with great difficulty that he performed the journey. His unequalled dignity and celebrity, however, added to the influence of powerful friends, commanded the respect of the tribunal before which he was summoned, which they manifested by permitting him to reside at the palace of his friend, the Tuscan Ambassador ; and when it became necessary, in the course of the inquiry, to examine him in person, although his removal to the Holy Office was then insisted upon, yet he was not, at first, like other heretics, committed to close and solitary confinement. On the contrary, he was lodged in the apartments of the Head of the Inquisition, where he was allowed the attendance of his own servant, who was also permitted to sleep in an adjoining room, and to come and go at pleasure. These were justly deemed extraordinary indulgences in an age when the punishment of heretics usually began before their trial. It has recently, however, been put beyond doubt, by the investigations of an eminent German scholar, that the aged philosopher did not submit to abjure the great truths he had established by his discourses, until he had been subjected to the horrid torture of the rack.

About four months after Galileo's arrival in Rome, he was summoned to the Holy Office. He was detained there during the whole of that day ; and on the next day was conducted, in a penitential dress, to the Convent of Minerva, where the Cardinals and Prelates, his judges, were assembled for the purpose of passing judgment upon him, by which this venerable old man was solemnly called upon to renounce and abjure, as impious and heretical, the opinions which his whole existence had been consecrated to form and strengthen. Probably there is not a more curious document, in the history of science, than that which contains the sentence of the Inquisition on Galileo, and his consequent abjuration. It teaches us so much, both of the darkness and bigotry of the terrible Inquisition, and of the sufferings encountered by those early martyrs of science, that I transcribe, from the excellent "Life of Galileo," in the "Library of Useful Knowledge," (from which I have borrowed much already,) the entire record of this transaction. The sentence of the Inquisition is as follows :—

"We, the undersigned, by the grace of God, Cardinals of the Holy Roman Church, Inquisitors General throughout the whole Christian Republic, Special Deputies of the Holy Apostolical Chair against heretical depravity :

"Whereas, you, Galileo, son of the late Vincenzo Galilei, of Florence, aged seventy years, were denounced, in 1615, to this Holy Office, for holding as true a false doctrine taught by many, namely, that the sun is immoveable in the centre of the world,. and that the earth moves, and also with a diurnal motion ; also, for having pupils which you instructed in the same opinions ; also, for maintaining a correspondence on the same with some German mathematicians ; also, for publishing certain letters on the solar spots, in which you developed the same doctrine as true ; also, for answering the objections which were continually produced from the Holy Scriptures, by glozing the said Scriptures, according to your own meaning ; and whereas, thereupon was produced the copy of a writing, in form of a letter, professedly written by you to a person formerly your pupil, in which following the hypothesis of Copernicus, you include several propositions contrary to the true sense and authority of the Holy Scriptures : therefore, this Holy Tribunal, being desirous of providing against the disorder and mischief which was thence proceeding and increasing, to the detriment of the holy faith, by the desire of His Holiness, and of the Most Eminent Lords Cardinals of this supreme and universal Inquisition, the two propositions of the stability of the sun, and motion of the earth, were *qualified* by the *Theological Qualifiers* as follows :

"1. The proposition that the sun is in the centre of the world, and immoveable from its place, is absurd, philosophically false, and formally heretical ; because it is expressly contrary to the Holy Scriptures.

"2. The proposition that the earth is not the centre of the world, nor immoveable, but that it moves, and also with a diurnal motion, is also absurd, philosophically false, and, theologically considered, equally erroneous in faith.

" But whereas, being pleased at that time to deal mildly with you, it was decreed in the Holy Congregation, held before His Holiness on the twenty-fifth day of February, 1616, that His Eminence the Lord Cardinal Bellarmine should enjoin you to give up altogether the said false doctrine ; if you should refuse, that you should be ordered by the Commissary of the Holy Office to relinquish it, not to teach it to others, nor to defend it, and in default of the acquiescence, that you should be imprisoned ; and in execution of this decree, on the following day, at the palace, in presence of His Eminence the said Lord Cardinal Bellarmine, after you had been mildly admonished by the said Lord Cardinal, you were commanded by the acting Commissary of the Holy Office, before a notary and witnesses, to relinquish altogether the said false opinion, and, in future, neither to defend nor teach it in any manner, neither verbally nor in writing, and upon your promising obedience, you were dismissed.

" And, in order that so pernicious a doctrine might be altogether rooted out, nor insinuate itself further to the heavy detriment of the Catholic truth, a decree emanated from the Holy Congregation of the Index, prohibiting the books which treat of this doctrine ; and it was declared false, and altogether contrary to the Holy and Divine Scripture.

" And whereas, a book has since appeared, published at Florence last year, the title of which showed that you were the author, which title is, ' The Dialogue of Galileo Galilei, on the two principal Systems of the World, the Ptolemaic and Copernican ; ' and whereas, the Holy Congregation has heard that, in consequence of printing the said book, the false opinion of the earth's motion and stability of the sun is daily gaining ground ; the said book has been taken into careful consideration, and in it has been detected a glaring violation of the said order, which had been intimated to you ; inasmuch as in this book you have defended the said opinion, already, and in your presence, condemned ; although, in the same book, you labour, with many circumlocutions, to induce the belief that it is left by you undecided, and in express terms probable ; which is equally a very grave error, since an opinion can in no way be probable which has been already declared and finally determined contrary to the Divine Scripture. Therefore, by Our order, you have been cited to this Holy Office, where, on your examination upon oath, you have acknowledged the said book as written and printed by you. You also confessed that you began to write the said book ten or twelve years ago, after the order aforesaid had been given. Also, that you demanded license to publish it, but without signifying to those who granted you this permission, that you had been commanded not to hold, defend, or teach, the said doctrine in any manner. You also confessed that the style of the said book was, in many places, so composed, that the reader might think the arguments adduced on the false side to be so worded, as more effectually to entangle the understanding than to be easily solved, alleging, in excuse, that you have thus run into an error, foreign (as you say) to your intention, from writing in the form of a dialogue, and in consequence of the natural complacency which every one feels with regard to his own subtilties, and in showing himself more skilful than the generality of mankind in contriving, even in favour of false propositions, ingenious and apparently probable arguments.

"And, upon a convenient time being given you for making your defence, you produced a certificate in the hand-writing of His Eminence, the Lord Cardinal Bellarmine, procured, as you said, by yourself, that you might defend yourself against the calumnies of your enemies, who reported that you had abjured your opinions, and had been punished by the Holy Office; in which certificate it is declared, that you had not abjured nor had been punished, but merely that the declaration made by His Holiness, and promulgated by the Holy Congregation of the Index, had been announced to you, which declares that the opinion of the motion of the earth and stability of the sun, is contrary to the Holy Scriptures, and therefore cannot be held or defended. Wherefore, since no mention is there made of two articles of the order, to wit, the order 'not to teach,' and 'in any manner,' you argued that we ought to believe that, in the lapse of fourteen or sixteen years, they had escaped your memory, and that this was also the reason why you were silent as to the order, when you sought permission to publish your book, and that this is said by you, not to excuse your error, but that it may be attributed to vain-glorious ambition rather than to malice. But this very certificate, produced on your behalf, has greatly aggravated your offence, since it is therein declared that the said opinion is contrary to the Holy Scriptures, and yet you have dared to treat of it, and to argue that it is probable; nor is there any extenuation in the license artfully and cunningly extorted by you, since you did not intimate the command imposed upon you. But whereas, it appeared to Us, that you had not disclosed the whole truth with regard to your intentions, We thought it necessary to proceed to the rigorous examination of you, in which (without any prejudice to what you had confessed, and which is above detailed against you, with regard to your said intention) you answered like a good Catholic.

" Therefore, having seen and maturely considered the merits of your cause, with your said confessions and excuses, and everything else which ought to be seen and considered, We have come to the underwritten final sentence against you :

" Invoking, therefore, the most holy name of our Lord Jesus Christ, and of his Most Glorious Virgin Mother, Mary, by this Our final sentence, which, sitting in council and judgment for the tribunal of the Reverend Masters of Sacred Theology, and Doctors of both Laws, Our Assessors, We put forth in this writing touching the matters and controversies before Us, beween the Magnificent Charles Sincerus, Doctor of both Laws, Fiscal Proctor of this Holy Office, of the one part, and you, Galileo Galilei, an examined and confessed criminal from this present writing now in progress, as above, of the other part, We pronounce, judge, and declare, that you, the said Galileo, by reason of these things which have been detailed in the course of this writing, and which, as above, you have confessed, have rendered yourself vehemently suspected, by this Holy Office, of heresy ; that is to say, that you believe and hold the false doctrine, and contrary to the Holy and Divine Scriptures, namely, that the sun is the centre of the world, and that it does not move from east to west, and that the earth does move, and is not the centre of the world ; also, that an opinion can be held and supported, as probable, after it has been declared and finally decreed contrary to the Holy Scripture, and, consequently, that you have incurred all the censures and penalties enjoined and promulgated in the sacred canons, and other general and particular constitutions against delinquents of this description. From which it is Our pleasure that you be absolved, provided that, with a sincere heart and unfeigned faith, in Our presence, you abjure, curse, and detest, the said errors and heresies, and every other error and heresy, contrary to the Catholic and Apostolic Church of Rome, in the form now shown to you.

" But that your grievous and pernicious error and transgression may not go altogether unpunished, and that you may be made more cautious in future, and may be a warning to others to abstain from delinquencies of this sort, We decree, that the book of the Dialogues of Galileo Galilei be prohibited by a public edict, and We condemn you to the formal prison of this Holy Office for a period determinable at Our pleasure ; and, by way of salutary penance, We order you, during the next three years, to recite, once a week, the seven penitential psalms, reserving to Ourselves the power of moderating, commuting, or taking off the whole or part of the said punishment or penance.

"And so We say, pronounce, and by Our sentence declare, decree, and reserve, in this and in every other better form and manner, which lawfully We may and can use.   So We, the subscribing Cardinals, pronounce." [Subscribed by seven Cardinals.]

In conformity with the foregoing sentence, Galileo was made to kneel before the Inquisition, and make the following *Abjuration :—*

" I, Galileo Galilei, son of the late Vincenzo Galilei, of Florence, aged seventy years, being brought personally to judgment, and kneeling before you, Most Eminent and Most Reverend Lords Cardinals, General Inquisitors of the Universal Christian Republic against heretical depravity, having before my eyes the Holy Gospels, which I touch with my own hands, swear, that I have always believed, and, with the help of God, will in future believe, every article which the Holy Catholic and Apostolic Church of Rome holds, teaches, and preaches.   But, because I had been enjoined, by this Holy Office, altogether to abandon the false opinion which maintains that the sun is the centre and immoveable, and forbidden to hold, defend, or teach, the said false doctrine, in any manner ; and after it had been signified to me that the said doctrine is repugnant to the Holy Scripture, I have written and printed a book, in which I treat of the same doctrine now condemned, and adduce reasons with great force in support of the same, without giving any solution, and therefore have been judged grievously suspected of heresy ; that is to say, that I held and believed that the sun is the centre of the world and immoveable, and that the earth is not the centre and moveable ; willing, therefore, to remove from the minds of Your Eminences, and of every Catholic Christian, this vehement suspicion rightfully entertained towards me, with a sincere heart and unfeigned faith, I abjure, curse, and detest, the said errors and heresies, and generally every other error and sect contrary to the said Holy Church ; and I swear, that I will never more in future say or assert anything, verbally or in writing, which may give rise to a similar suspicion of me ; but if I shall know any heretic, or any one suspected of heresy, that I will denounce him to this Holy Office, or to the Inquisitor and Ordinary of the place in which I may be. I swear, moreover, and promise, that I will fulfil and observe fully, all the penances which have been or shall be laid on me by this Holy Office. But if it shall happen that I violate any of my said promises, oaths, and protestations (which God avert !) I subject myself to all the pains and punishments which have been decreed and promulgated by the sacred canons, and other general and particular constitutions, against delinquents of this description. So may God help me, and his Holy Gospels, which I touch with my own hands. I, the above-named Galileo Galilei, have abjured, sworn, promised, and bound myself as above ; and, in witness thereof, with my own hand have subscribed this present writing of my abjuration, which I have recited, word for word.

" At Rome, in the Convent of Minerva, twenty-second of June 1633, I, Galileo Galilei, have abjured as above, with my own hand."

Rising from his knees, after this humiliating lie, wrung from the aged philosopher by such horrible means, he turned to one who stood near, and exclaimed, " *E pur se muove.*" For all this, it does move !

From the court Galileo was conducted to prison, to be immured for life in one of the dungeons of the Inquisition. His sentence was afterwards mitigated, and he was permitted to return to Florence ; but the humiliation to which he had been subjected pressed even more heavily on his spirits than the cruel tortures to which he had been subjected, beset as he already was with infirmities, and totally blind, and he never more talked or wrote on the subject of astronomy.

There is enough in the character of Galileo to command a high admiration. There is much, also, in his sufferings in the cause of science, to excite the deepest sympathy, and even compassion. He is, moreover, universally represented to have been a man of great equanimity, and of a noble and generous disposition. No scientific character of the age, or perhaps of any age, forms a structure of finer proportions, or wears in a higher degree the grace of symmetry. Still we cannot but regret his employing artifice in the promulgation of truth ; and we are compelled to lament that his lofty spirit bowed in the final conflict. How far, therefore, he sinks below the dignity of the Christian martyr ! "At the age of seventy," says Dr. Brewster, in his life of Sir Isaac Newton, "on his bended knees, and with his right hand resting on the Holy Evangelists, did this patriarch of science avow his present and past belief in the dogmas of the Romish Church, abandon as false and heretical the doctrine of the earth's motion and of the sun's immobility, and pledge himself to denounce to the Inquisition any other person who was even suspected of heresy. He abjured, cursed, and detested, those eternal and immutable truths which the Almighty had permitted him to be the first to establish. Had Galileo but added the courage of the martyr to the wisdom of the sage ; had he carried the glance of his indignant eye round the circle of his judges ; had he lifted his hands to heaven, and called the living God to witness the truth and immutability of his opinions, the bigotry of his enemies would have been disarmed, and science would have enjoyed a memorable triumph."

KEPLER.

KEPLER was a native of Germany. He was born in the Duchy of Wurtemberg, in 1571. As Copernicus, Tycho Brahe, Galileo, Kepler, and Newton, are names that are much associated in the history of astronomy, let us see how they stood related to each other in point of time. Copernicus was born in 1473 : Tycho, in 1546 ; Galileo, in 1564 ; Kepler, in 1571 ; and Newton, in 1642. Hence, Copernicus was seventy-three years before Tycho, and Tycho ninety-six years before Newton. They all lived to an advanced age, so that Tycho, Galileo, and Kepler, were contemporary for many years ; and Newton was born the year that Galileo died.

Kepler was born of parents who were then in humble circumstances, although of noble descent. Their misfortunes, which had reduced them to poverty, seem to have been aggravated by their own unhappy dispositions ; for his biographer informs us, that "his mother was treated with a degree of barbarity by her husband and brother-in-law, that was hardly exceeded by her own perverseness." It is fortunate, therefore, that Kepler, in his childhood, was removed from the immediate society and example of his parents, and educated at a public school at the expense of the Duke of Wurtemberg. He early imbibed a taste for natural philosophy, but had conceived a strong prejudice against astronomy, and even a contempt for it, inspired, probably, by the arrogant and ridiculous pretensions of the astrologers. who constituted the principal astronomers of his country. A vacant post, however, of teacher of astronomy, occurred when he was of a suitable age to fill it, and he was compelled to take it by the authority of his tutors, though with many protestations on his part, wishing to be provided for in some other more brilliant profession.

Happy is genius, when it lights on a profession entirely consonant to its powers, where the objects successively presented to it are so exactly suited to its nature, that it clings to them as the loadstone to its kindred metal among piles of foreign ores. Nothing could have been more congenial to the very mental constitution of Kepler, than the study of astronomy,—a science where the most capacious understanding may find scope in unison with the most fervid imagination.

Much as has been said against hypotheses in philosophy, it is nevertheless, a fact, that some of the greatest truths have been discovered in the pursuit of hypotheses, in themselves entirely false; truths, moreover, far more important than those assumed by the hypotheses; as Columbus, in searching for a north-west passage to India, discovered a new world. Thus Kepler groped his way through many false and absurd suppositions, to some of the most sublime discoveries ever made by man. The fundamental principle which guided him was not, however, either false or absurd. It was that God, who made the world, had established, throughout all his works, fixed laws,—laws that are often so definite as to be capable of expression in exact numerical terms. In accordance with these views, he sought for numerical relations in the disposition and arrangement of the planets, in respect to their number, the times of their revolution, and their distances from one another. Many, indeed, of the subordinate suppositions which he made, were extremely fanciful; but he tried his own hypotheses by a rigorous mathematical test, wherever he could apply it; and as soon as he discovered that a supposition would not abide this test, he abandoned it without the least hesitation, and adopted others, which he submitted to the same severe trial, to share, perhaps, the same fate. "After many failures," he says, "I was comforted by observing that the motions, in every case, seemed to be connected with the distances; and that, when there was a great gap between the orbits, there was the same between the motions. And I reasoned that, if God·had adapted motions to the orbits in some relation to the distances, he had also arranged the distances themselves in relation to something else."

In two years after he commenced the study of astronomy, he published a book called the "*Mysterium Cosmographicum*," a name which implies an explanation of the mysteries involved in the construction of the universe. This work was full of the wildest speculations and most extravagant hypotheses, the most remarkable of which was, that the distances of the planets from the sun are regulated by the relations which subsist between the five regular solids. It is well known to geometers that there are, and can be, only five *regular solids*. These are, first, the *tetraedron*, a four-sided figure, all whose sides are equal and similar triangles; secondly, the *cube*, contained by six equal squares; thirdly, an *octaedron*, an eight-sided figure, consisting of two four-sided pyramids joined at their bases; fourthly, a *dodecaedron*, having twelve five-sided, or pentagonal faces; and, fifthly, an *icosaedron*, contained by twenty equal and similar triangles. · You will be much at a loss, I think, to imagine what relation Kepler could trace between these strange figures and the distances of the several planets from the sun. He thought he discovered a connexion between those distances and the spaces which figures of this kind would occupy, if interposed in certain ways between them. Thus, he says, the Earth is a circle, the measure of all; round it describe a dodecaedron, and the circle including this will be the orbit of Mars. Round this circle describe a tetraedron, and the circle including this will be the orbit of Jupiter. Describe a cube round this, and the circle including it will be the orbit of Saturn. Now, inscribe in the earth an icosaedron, and the circle included in this will give the orbit of Venus. In this inscribe an octaedron, and the circle included in this will be the orbit of Mercury. On this supposed discovery Kepler exults in the most enthusiastic expressions. " The intense pleasure I have received from this discovery never can be told in words. I regretted no more time wasted; I tired of no labour; I shunned no toil of reckoning; days and nights I spent in calculations, until I could see whether this opinion would agree with the orbits of Copernicus, or whether my joy was to vanish into air. I willingly subjoin that sentiment of Archytas, as given by Cicero : ' If I could mount up into heaven, and thoroughly perceive the nature of the world and the beauty of the stars, that admiration would be without a charm for me, unless I had some one like you, reader, candid, attentive, and eager for knowledge, to whom to describe it.' If you acknowledge this feeling, and are candid, you will refrain from blame, such as, not without cause, I anticipate : but if, leaving that to itself, you fear, lest these things be not ascertained, and that I have shouted triumph before victory, at least approach these pages, and learn the matter in consideration : you will not find, as just now, new and unknown planets interposed ; that boldness of mine is not approved ; but those old ones very little loosened, and so furnished by the interposition (however absurd you may think it) of rectilinear figures, that in future you may give a reason to the rustics, when they ask for the hooks which keep the skies from falling."

When Tycho Brahe, who had then retired from his famous Uraniburg, and was settled in Prague, met with this work of Kepler's, he immediately recognised under this fantastic garb the lineaments of a great astronomer. He needed such an unwearied and patient calculator as he perceived Kepler to be, to aid him in his labours, in order that he might devote himself more unreservedly to the taking of observations,—an employment in which he delighted, and in which, as I mentioned in giving you a sketch of his history, he excelled all men of that and preceding ages. Kepler, therefore, at the express invitation of Tycho, went to Prague, and joined him in the capacity of assistant. Had Tycho been of a nature less truly noble, he might have looked with contempt on one who had made so few observations, and indulged so much in wild speculation ; or he might have been jealous of a rising genius, in which he descried so many signs of future eminence as an astronomer ; but, superior to all the baser motives, he extends to the young aspirant the hand of encouragement in the following kind invitation : " Come, not as a stranger, but as a very welcome friend ; come, and share in my observations, with such instruments as I have with me."

Several years previous to this, Kepler, after one or two unsuccessful trials, had found him a wife, from whom he expected a considerable fortune ; but in this he was disappointed ; and so poor was he, that, when on his journey to Prague in company with his wife, being taken sick, he was unable to defray the expenses of the journey, and was forced to cast himself on the bounty of Tycho.

In the course of the following year, while absent from Prague, he fancied that Tycho had injured him, and accordingly addressed to the noble Dane a letter full of insults and reproaches. A mild reply from Tycho opened the eyes of Kepler to his own ingratitude. His better feelings soon returned, and he sent to his great patron this humble apology : "Most noble Tycho !

How shall I enumerate, or rightly estimate, your benefits conferred on me! For two months you have liberally and gratuitously maintained me, and my whole family ; you have provided for all my wishes ; you have done me every possible kindness ; you have communicated to me every thing you hold most dear ; no one, by word or deed, has intentionally injured me in anything ; in short, not to your own children, your wife, or yourself, have you shown more indulgence than to me. This being so, as I am anxious to put upon record, I cannot reflect, without consternation, that I should have been so given up by God to my own intemperance, as to shut my eyes on all these benefits ; that, instead of modest and respectful gratitude, I should indulge for three weeks in continual moroseness towards all your family, and in headlong passion and the utmost insolence towards yourself, who possess so many claims on my veneration, from your noble family, your extraordinary learning, and distinguished reputation. Whatever I have said or written against the person, the fame, the honour, and the learning of your Excellency ; or whatever, in any other way, I have injuriously spoken or written (if they admit no other more favourable interpretation), as, to my grief, I have spoken and written many things, and more than I can remember ; all and everything I recant, and freely and honestly declare and profess to be groundless, false, and incapable of proof." This was ample satisfaction to the generous Tycho.

"To err is human: to forgive, divine."

On Kepler's return to Prague, he was presented to the Emperor by Tycho, and honoured with the title of Imperial Mathematician. This was in 1601, when he was thirty years of age. Tycho died shortly after, and Kepler succeeded him as principal mathematician to the Emperor ; but his salary was badly paid, and he suffered much from pecuniary embarrassments. Although he held the astrologers, or those who told fortunes by the stars, in great contempt, yet he entertained notions of his own, on the same subject, quite as extravagant, and practised the art of casting nativities to eke out a support for his family.

When Galileo began to observe with his telescope, and announced, in rapid succession, his wonderful discoveries, Kepler entered into them with his characteristic enthusiasm, although

T

they subverted many of his favourite hypotheses. But such was his love of truth, that he was among the first to congratulate Galileo, and a most engaging correspondence was carried on between these master-spirits.

The first planet which occupied the particular attention of Kepler was Mars, the long and assiduous study of whose motions conducted him at length to the discovery of those great principles called "Kepler's Laws." Rarely do we meet with so remarkable a union of a vivid fancy with a profound intellect. The hasty and extravagant suggestions of the former were submitted to the most laborious calculations, some of which, that were of great length, he repeated seventy times. This exuberance of fancy frequently appears in his style of writing, which occasionally assumes a tone ludicrously figurative. He seems constantly to contemplate Mars as a valiant hero, who had hitherto proved invincible, and who would often elude his own efforts to conquer him. "While thus triumphing over Mars, and preparing for him, as for one altogether vanquished, tabular prisons, and equated, eccentric fetters, it is buzzed here and there that the victory is vain, and that the war is raging anew as violently as before. For the enemy, left at home a despised captive, has burst all the chains of the equation, and broken forth of the prisons of the tables. Skirmishes routed my forces of physical causes, and shaking off the yoke, regained their liberty. And now there was little to prevent the fugitive enemy from effecting a junction with his own rebellious supporters, and reducing me to despair, had I not suddenly sent into the field a reserve of new physical reasonings on the rout and dispersion of the veterans, and diligently followed, without allowing the slightest respite, in the direction in which he had broken out."

But he pursued this warfare with the planet until he gained a full conquest, by the discovery of the first two of his laws; namely, that *he revolves in an elliptical orbit*, and that *his radius vector passes over equal spaces in equal times.*

Domestic troubles, however, involved him in the deepest affliction. Poverty, the loss of a promising and favourite son, the death of his wife, after a long illness;—these were some of the misfortunes that clustered around him. Although his first marriage had been an unhappy one, it was not consonant to his genius to surrender anything after only a single trial. Accordingly, it was not long before he contracted a second alliance.

He commissioned a number of his friends to look out for him, and he soon obtained a tabular list of eleven names, among whom his affections wavered. The progress of his courtship is thus narrated in the interesting "Life" contained in the "Library of Useful Knowledge." It furnishes so fine a specimen of his eccentricities, that I cannot deny myself the pleasure of transcribing the passage. It is taken from an account which Kepler himself gave in a letter to a friend.

"The first on the list was a widow, an intimate friend of his first wife, and who, on many accounts, appeared a most eligible match. At first, she seemed favourably inclined to the proposal; it is certain that she took time to consider it, but at last she very quietly excused herself. Finding her afterwards less agreeable in person than he had anticipated, he considered it a fortunate escape, mentioning, among other objections, that she had two marriageable daughters, whom, by the way, he had got on his list for examination. He was much troubled to reconcile his astrology with the fact of his having taken so much pains about a negotiation not destined to succeed. He examined the case professionally. 'Have the stars,' says he, 'exercised any influence here? For just about this time, the direction of the mid-heaven is in hot opposition to Mars, and the passage of Saturn through the ascending point of the zodiac, in the scheme of my nativity, will happen again next November and December. But, if these are the causes, how do they act? Is that explanation the true one, which I have elsewhere given? For I can never think of handing over to the stars the office of deities, to produce effects. Let us, therefore, suppose it accounted for by the stars, that at this season I am violent in my temper and affections, in rashness of belief, in a show of pitiful tender-heartedness, in catching at reputation by new and paradoxical notions, and the singularity of my actions; in busily inquiring into, and weighing, and discussing various reasons; in the uneasiness of my mind, with respect to my choice. I thank God that that did not happen which might have happened; that this marriage did not take place. Now for the others.' Of these, one was too old; another, in bad health; another, too proud of her birth and quarterings; a fourth had learned nothing but showy accomplishments, not at all suitable to the kind of life she would have to lead with him. Another grew impatient, and married a more decided admirer while he was hesitating. 'The mischief,' says he, 'in all these

T 2

attachments was, that, whilst I was delaying, comparing, and
balancing conflicting reasons, every day saw me inflamed with a
new passion.' By the time he reached No. 8 of his list, he found
his match in this respect. 'Fortune has avenged herself at length
on my doubtful inclinations. At first, she was quite complying,
and her friends also. Presently, whether she did or did not
consent, not only I, but she herself did not know. After the
lapse of a few days came a renewed promise, which, however,
had to be confirmed a third time : and, four days after that, she
again repented her confirmation, and begged to be excused from
it. Upon this, I gave her up, and this time all my counsellors
were of one opinion.' This was the longest courtship in the list,
having lasted three whole months ; and, quite disheartened by its
bad success, Kepler's next attempt was of a more timid com-
plexion. His advances to No. 9 were made by confiding to her
the whole story of his recent disappointment, prudently determining to be guided in his behaviour by observing whether the
treatment he experienced met with a proper degree of sympathy.
Apparently, the experiment did not succeed ; and, when almost
reduced to despair, Kepler betook himself to the advice of a
friend, who had for some time past complained that she was not
consulted in this difficult negotiation. When she produced
No. 10, and the first visit was paid, the report upon her was as
follows: 'She has, undoubtedly, a good fortune, is of good family,
and of economical habits : but her physiognomy is most horribly
ugly ; she would be stared at in the streets, not to mention the
striking disproportion in our figures. I am lank, lean, and spare ;
she is short and thick. In a family notorious for fatness, she is
considered superfluously fat.' The only objection to No. 11 seems
to have been her excessive youth ; and when this treaty was
broken off on that account, Kepler turned his back upon all his
advisers, and chose for himself one who had figured as No. 5 in
his list, to whom he professes to have felt attached throughout,
but from whom the representations of his friends had hitherto
detained him, probably on account of her humble station."

Having thus settled his domestic affairs, Kepler now betook
himself, with his usual industry, to his astronomical studies, and
brought before the world the most celebrated of his publications,
entitled, "Harmonics." In the fifth book of this work he
announced his *Third Law*,—that the squares of the periodical
times of the planets are as the cubes of the distances. Kepler's

rapture upon detecting it was unbounded. "What," says he, "I prophesied two-and-twenty years ago, as soon as I discovered the five solids among the heavenly orbits ; what I firmly believed long before I had seen Ptolemy's Harmonics ; what I had promised my friends in the title of this book, which I named before I was sure of my discovery ; what sixteen years ago I urged as a thing to be sought ; that for which I joined Tycho Brahe, for which I settled in Prague, for which I have devoted the best part of my life to astronomical contemplations ;—at length I have brought to light, and have recognised its truth beyond my most sanguine expectations. It is now eighteen months since I got the first glimpse of light, three months since the dawn, very few days since the unveiled sun, most admirable to gaze on, burst out upon me. Nothing holds me : I will indulge in my sacred fury ; I will triumph over mankind by the honest confession that I have stolen the golden vases of the Egyptians to build up a tabernacle for my God far from the confines of Egypt. If you forgive me, I rejoice : if you are angry, I can bear it ; the die is cast, the book is written, to be read either now or by posterity, I care not which. I may well wait a century for a reader, as God has waited six thousand years for an observer." In accordance with the notion he entertained respecting the "music of the spheres," he made Saturn and Jupiter take the bass, Mars the tenor, the Earth and Venus the counter, and Mercury the treble.

"The misery in which Kepler lived," says Sir David Brewster, in his "Life of Newton," "forms a painful contrast with the services which he performed for science. The pension on which he subsisted was always in arrears ; and though the three emperors, whose reigns he adorned, directed their ministers to be more punctual in its payment, the disobedience of their commands was a source of continual vexation to Kepler. When he retired to Silesia, to spend the remainder of his days, his pecuniary difficulties became still more harassing. Necessity at length compelled him to apply personally for the arrears which were due ; and he accordingly set out, in 1630, when nearly sixty years of age, for Ratisbon ; but, in consequence of the great fatigue which so long a journey on horseback produced, he was seized with a fever, which put an end to his life."

Professor Whewell (in his interesting work on Astronomy and General Physics considered with reference to natural Theology) expresses the opinion that Kepler, notwithstanding his constitutional oddities, was a man of strong and lively piety. His "Commentaries on the motions of Mars" he opens with the following passage : "I beseech my reader, that, not unmindful of the Divine goodness bestowed on man, he do with me praise and celebrate the wisdom and greatness of the Creator, which I open to him from a more inward explication of the form of the world, from a searching of causes, from a detection of the errors of vision ; and that thus, not only in the firmness and stability of the earth, he perceive with gratitude the preservation of all living things in nature as the gift of God, but also that in its motion, so recondite, so admirable, he acknowledge the wisdom of the Creator. But him who is too dull to receive this science, or too weak to believe the Copernican system without harm to his piety;—him, I say, I advise that leaving the school of astronomy, and condemning, if he please, any doctrines of the philosophers, he follow his own path, and desist from this wandering through the universe ; and, lifting up his natural eyes, with which he alone can see, pour himself out in his own heart, in praise of God the Creator ; being certain that he gives no less worship to God than the astronomer, to whom God has given to see more clearly with his inward eye, and who, for what he has himself discovered, both can and will glorify God."

In a Life of Kepler, very recently published in his native country, founded on manuscripts of his which have lately been brought to light, there are given numerous other examples of a similar devotional spirit. Kepler thus concludes his Harmonics : "I give Thee thanks, Lord and Creator, that Thou hast given me joy through Thy creation ; for I have been ravished with the work of Thy hands. I have revealed unto mankind the glory of Thy works, as far as my limited spirit could conceive their infinitude. Should I have brought forward anything that is unworthy of Thee, or should I have sought my own fame, be graciously pleased to forgive me."

As Galileo experienced the most bitter persecutions from the Church of Rome, so Kepler met with much violent opposition and calumny from the Protestant clergy of his own country, particularly for adopting, in an almanac which, as astronomer royal, he annually published, the reformed calendar, as given by the Pope of Rome. His opinions respecting religious liberty, also, appear to have been greatly in advance of the times in which he lived. In answer to certain calumnies with which he was assailed, for his boldness in reasoning from the light of nature, he uttered these memorable words: "The day will soon break, when pious simplicity will be ashamed of its blind super-stition ; when men will recognise truth in the book of nature as well as in the Holy Scriptures, and rejoice in the two revelations."

NEWTON.

Sir Isaac Newton was born in Lincolnshire, in 1642, just one year after the death of Galileo. His father died before he was born, and he was a helpless infant, of a diminutive size, and so feeble a frame, that his attendants hardly expected his life for a single hour. The family dwelling was of humble architecture, situated in a retired but beautiful valley, and was surrounded by a small farm, which afforded but a scanty living to the widowed mother and her precious charge. It will pro-bably be found that genius has oftener emanated from the cottage than from the palace.

The boyhood of Newton was distinguished chiefly for his ingenious mechanical contrivances. Among other pieces of mechanism, he constructed a windmill, so curious and complete in its workmanship, as to excite universal admiration. After carrying it a while by the force of the wind, he resolved to sub-stitute animal power ; and for this purpose he inclosed in it a mouse, which he called the miller, and which kept the mill going by acting on a tread-wheel. The power of the mouse was brought into action by unavailing attempts to reach a portion of corn placed above the wheel. A water-clock, a four-wheeled carriage propelled by the rider himself, and kites of superior workmanship, were among the productions of the mechanical genius of this gifted boy. At a little later period he began to turn his attention to the motions of the heavenly bodies, and constructed several sun-dials on the walls of the house where he lived. All this was before he had reached his fifteenth year. At this age, he was sent by his mother, in company with an old family servant, to a neighbouring market-town, to dispose of products of their farm, and to buy articles of merchandise for

their family use ; but the young philosopher left all these nego-
tiations to his worthy partner, occupying himself meanwhile
with a collection of old books, which he had found in a garret. At
other times he stopped on the road, and took shelter with his book
under a hedge, until the servant returned. They endeavoured
to educate him as a farmer ; but the perusal of a book, the construction of a water-mill, or some other mechanical or scientific
amusement, absorbed all his thoughts, when the sheep were
going astray, and the cattle were devouring or treading down
the corn. One of his uncles having found him one day under a
hedge, with a book in his hand, and entirely absorbed in meditation, took it from him, and found that it was a mathematical
problem which so engrossed his attention. His friends, therefore,
wisely resolved to favour the bent of his genius, and removed
him from the farm to school, to prepare for the university. In
the eighteenth year of his age, Newton was admitted to Trinity
College, Cambridge. He made rapid and extraordinary advances
in the mathematics, and soon afforded unequivocal presages of
that greatness which afterwards placed him foremost among the
master spirits of the world. In 1669, at the age of twenty-seven,
he became professor of mathematics at Cambridge, a post which
he occupied for many years afterwards. During the four or five
years previous to this he had, in fact, made most of those great
discoveries which have immortalised his name.

ANCIENT AND MODERN IDEAS

ON

THE SYSTEM OF THE WORLD.

—•—

By a system of the world, I understand an explanation of *the arrangement of all the bodies that compose the material universe, and of their relations to each other.* It is otherwise called the "Mechanism of the Heavens;" and indeed, in the system of the world, we figure to ourselves a machine, all parts of which have a mutual dependence, and conspire to one great end. "The machines that were first invented," says Adam Smith, "to perform any particular movement, are always the most complex; and succeeding artists generally discover that, with fewer wheels, and with fewer principles of motion, than had originally been employed, the same effects may be more easily produced. The first systems, in the same manner, are always the most complex; and a particular connecting chain or principle is generally thought necessary, to unite every two seemingly disjointed appearances; but it often happens, that *one great connecting principle* is afterwards found to be sufficient to bind together all the discordant phenomena that occur in a whole species of things!" This remark is strikingly applicable to the origin and progress of systems of astronomy. It is a remarkable fact in the history of the human mind, that astronomy is the oldest of the sciences, having been cultivated, with no small success, long before any attention was paid to the causes of the common terrestrial phenomena. The opinion has always prevailed among those who were unenlightened by science, that very extraordinary appearances in the sky, as comets, fiery meteors, and eclipses, are omens of the wrath of heaven. They have, therefore, in all ages, been watched with the greatest attention; and their appearances have been minutely recorded by the historians of the times. The idea, moreover, that the aspects of the stars are connected with the destinies of individuals and of empires, has been remarkably prevalent from the earliest records of history down to a very late period, and, indeed, still lingers among the uneducated and credulous. This notion gave rise to ASTROLOGY, —an art which professed to be able, by a knowledge of the varying aspects of the planets and stars, to penetrate the veil of futurity, and to foretell approaching irregularities of nature herself, and the fortunes of kingdoms and of individuals. That department of astrology which took cognisance of extraordinary occurrences in the natural world, as tempests, earthquakes, eclipses, and volcanoes, both to predict their approach and to interpret their meaning, was called *natural astrology;* that which related to the fortunes of men and of empires, *judicial astrology.* Among many ancient nations, astrologers were held in the highest estimation, and were kept near the persons of monarchs; and the practice of the art constituted a lucrative profession throughout the middle ages. Nor were the ignorant and un-educated portions of society alone the dupes of its pretensions. Hippocrates, the "Father of Medicine," ranks astrology among the most important branches of knowledge to the physician; and Tycho Brahe and Lord Bacon were firm believers in its mysteries. Astrology, fallacious as it was, must be acknowledged to have rendered the greatest services to astronomy, by leading to the accurate observation and diligent study of the stars.

At a period of very remote antiquity, astronomy was cultivated in China, India, Chaldea, and Egypt. The Chaldeans were par-ticularly distinguished for the accuracy and extent of their astronomical observations. Calisthenes, the Greek philosopher, who accompanied Alexander the Great in his Eastern conquests, transmitted to Aristotle a series of observations made at Babylon nineteen centuries before the capture of that city by Alexander; and the wise men of Babylon and the Chaldean astrologers are referred to in the Sacred Writings. They enjoyed a clear sky and a mild climate, and their pursuits as shepherds favoured long-continued observations; while the admiration and respect accorded to the profession, rendered it an object of still higher ambition.

In the seventh century before the Christian era, astronomy began to be cultivated in Greece; and there arose successively three celebrated astronomical schools,—the school of Miletus, the school of Crotona, and the school of Alexandria. The first was established by Thales, six hundred and forty years before Christ; the second, by Pythagoras, one hundred and forty years afterwards; and the third, by the Ptolemies of Egypt, about three hundred years before the Christian era. As Egypt and Babylon were renowned among the most ancient nations, for their knowledge of the sciences, long before they were cultivated in Greece, it was the practice of the Greeks, when they aspired to the character of philosophers and sages, to resort to these countries to imbibe wisdom at its fountains. Thales, after extensive travels in Crete and Egypt, returned to his native place, Miletus, a town on the coast of Asia Minor, where he established the first school of astronomy in Greece. Although the minds of these ancient astronomers were beclouded with much error, yet Thales taught a few truths which do honour to his sagacity. He held that the stars are formed of fire; that the moon receives her light from the sun, and is invisible at her conjunctions because she is hid in the sun's rays. He taught the sphericity of the earth, but adopted the common error of placing it in the centre of the world. He introduced the division of the sphere into five zones, and taught the obliquity of the ecliptic. He was acquainted with the Saros, or sacred period of the Chaldeans, and employed it in calculating eclipses. It was Thales that predicted the famous eclipse of the sun which terminated the war between the Lydians and the Medes. Indeed, Thales is universally regarded as a bright but solitary star, glimmering through mists on the distant horizon.

To Thales succeeded, in the school of Miletus, two other astronomers of much celebrity, Anaximander and Anaxagoras. Among many absurd things held by Anaximander, he first taught the sublime doctrine that the planets are inhabited, and that the stars are suns of other systems. Anaxagoras attempted to explain all the secrets of the skies by natural causes. His reasonings, indeed, were alloyed with many absurd notions; but still he alone, among the astronomers, maintained the existence of one God. His doctrines alarmed his countrymen, by their audacity and impiety to their gods, whose prerogatives he was thought to invade; and, to deprecate their wrath, sentence of death was pronounced on the philosopher and all his family,—a sentence which was commuted only for the sad alternative of perpetual banishment. The very genius of the heathen mythology was at war with the truth. False in itself, it trained the mind to the love of what was false in the interpretation of nature; it arrayed itself against the simplicity of truth, and persecuted and put to death its most ardent votaries. The religion of the Bible, on the other hand, lends all its aid to truth in nature as well as in morals and religion. In its very genius it inculcates and inspires the love of truth; it suggests, by its analogies, the existence of established laws in the system of the world; and holds out the moon and stars, which the Creator has ordained, as fit objects to give us exalted views of his glory and wisdom.

Pythagoras was the founder of the celebrated school of Crotona. He was a native of Samos, an island in the Ægean sea, and flourished about five hundred years before the Christian era. After travelling more than thirty years in Egypt and Chaldea, and spending several years more at Sparta, to learn the laws and institutions of Lycurgus, he returned to his native island to dispense the riches he had acquired to his countrymen. But they, probably fearful of incurring the displeasure of the gods by the freedom with which he inquired into the secrets of the skies, gave him so unwelcome a reception, that he retired from them, in disgust, and established his school at Crotona, on the south-eastern coast of Italy. Hither, as to an oracle, the fame of his wisdom attracted hundreds of admiring pupils, whom he instructed in every species of knowledge. From the visionary notions which are generally understood to have been entertained on the subject of astronomy, by the ancients, we are apt to imagine that they knew less than they actually did of the truths of this science. But Pythagoras was acquainted with many important facts in astronomy, and entertained many opinions respecting the system of the world, which are now held to be true. Among other things well known to Pythagoras, either derived from his own investigations, or received from his predecessors, were the following; and we may note them as a synopsis of the state of astronomical knowledge at that age of the world. First, the principal *constellations*. These had begun to be formed in the earliest ages of the world. Several of them, bearing the same name as at present, are mentioned in the writings of Hesiod and Homer; and the " sweet influences of the Pleiades," and the " bands of Orion," are beautifully alluded to in the book of Job. Secondly, *eclipses*. Pythagoras knew both the causes of eclipses and how to predict them; not, indeed, in the accurate manner now practised, but by means of the Saros. Thirdly, Pythagoras had divined the true *system of the world*, holding that the sun, and not the earth, (as was generally held by the ancients, even for many ages after Pythagoras,) is the centre around which all the planets revolve ; and that the stars are so many suns, each the centre of a system like our own. Among lesser things, he knew that the earth is round ; that its surface is naturally divided into five zones ; and that the ecliptic is inclined to the equator. He also held that the earth revolves daily on its axis, and yearly around the sun ; that the galaxy is an assemblage of small stars ; and that it is the same luminary, namely, Venus, that constitutes both the morning and evening star ; whereas all the ancients before him had supposed that each was a separate planet, and accordingly the morning star was called Lucifer, and the evening star, Hesperus. He held, also, that the planets were inhabited, and even went so far as to calculate the size of some of the animals in the moon. Pythagoras was also so great an enthusiast in music, that he not only assigned to it a conspicuous place in his system of education, but even supposed that the heavenly bodies themselves were arranged at distances corresponding to the intervals of the diatonic scale, and imagined them to pursue their sublime march to notes created by their own harmonious movements, called the "music of the spheres ;" but he maintained that this celestial concert, though loud and grand, is not audible to the feeble organs of man, but only to the gods. With few exceptions, however, the opinions of Pythagoras on the system of the world were founded in truth. Yet they were rejected by Aristotle, and by most succeeding astronomers, down to the time of Copernicus ; and in their place was substituted the doctrine of *crystalline spheres*, first taught by Eudoxus, who lived about three hundred and seventy years before Christ. According to this system, the heavenly bodies are set like gems in hollow solid orbs, composed of crystal so transparent, that no anterior orb obstructs in the least the view of any of the orbs that lie behind it. The sun and the planets have each its separate orb ; but the fixed stars are all set in the same grand orb ; and beyond this is another still, the *primum mobile*, which revolves daily, from east to west, and carries along with it all the other orbs. Above the whole spreads the *grand empyrean*, or third heavens, the abode of perpetual serenity.

To account for the planetary motions, it was supposed that each of the planetary orbs, as well as that of the sun, has a motion of its own eastward, while it partakes of the common diurnal motion of the starry sphere. Aristotle taught that these motions are effected by a tutelary genius of each planet, residing in it, and directing its motions, as the mind of man directs his movements.

Two hundred years after Pythagoras, arose the famous school of Alexandria, under the Ptolemies. These were a succession of Egyptian kings, and are not to be confounded with Ptolemy the astronomer. By the munificent patronage of this enlightened family for the space of three hundred years, beginning at the death of Alexander the Great, from whom the eldest of the Ptolemies had received his kingdom, the school of Alexandria concentrated in its vast library and princely halls, erected for the accommodation of the philosophers, nearly all the science and learning of the world. In wandering over the immense territories of ignorance and barbarism which covered, at the time, almost the entire face of the earth, the eye reposes upon this little spot as upon a verdant island in the midst of the desert. Among the choice fruits that grew in this garden of astronomy were several of the most distinguished ornaments of ancient science, of whom the most eminent were Hipparchus and Ptolemy. Hipparchus is justly considered as the Newton of antiquity. He sought his knowledge of the heavenly bodies not in the illusory suggestions of a fervid imagination, but in the vigorous application of an intellect of the first order. Previous to this period, celestial observations were made with the naked eye: but Hipparchus was in possession of instruments for measuring angles, and knew how to resolve spherical triangles. These were great steps beyond all his predecessors. He ascertained the length of the year within six minutes of the truth. He discovered the eccentricity, or elliptical figure, of the solar orbit, although he supposed the sun actually to move uniformly in a circle, but the earth to be placed out of the centre. He also determined the positions of the points among the stars where the earth is nearest the sun, and where it is most remote from it. He formed very accurate estimates of the obliquity of the ecliptic, and of the precession of the equinoxes. He computed the exact period of the synodic revolution of the moon, and the inclination of the lunar orbit; discovered the backward motion of her node and of her line of apsides; and made the first attempts to ascertain the horizontal parallaxes of the sun and moon. Upon the appearance of a new star in the firmament, he undertook, as already mentioned, to number the stars, and to assign to each its true place in the heavens, in order that posterity might have the means of judging what changes, if any, were going forward among these apparently unalterable bodies.

Although Hipparchus is generally considered as belonging to the Alexandrian school, yet he lived at Rhodes, and there made his astronomical observations, about one hundred and forty years before the Christian era. One of his treatises has come down to us; but his principal discoveries have been transmitted through the "Almagest" of Ptolemy. Ptolemy flourished at Alexandria nearly three centuries after Hipparchus, in the second century after Christ. His great work, the "Almagest," which has conveyed to us most that we know respecting the astronomical knowledge of the ancients, was the universal text-book of astronomers for fourteen centuries.

The name of this celebrated astronomer has also descended to us, associated with the system of the world which prevailed from Ptolemy to Copernicus, called the *Ptolemaic System*. The doctrines of the Ptolemaic system did not originate with Ptolemy, but, being digested by him out of materials furnished by various hands, it has come down to us under the sanction of his name. According to this system, the earth is the centre of the universe, and all the heavenly bodies daily revolve around it, from east to west. But although this hypothesis would account for the apparent diurnal motion of the firmament, yet it would not account for the apparent annual motion of the sun, nor for the slow motions of the planets from west to east. In order to explain these phenomena, recourse was had to *deferents* and *epicycles*,— an explanation devised by Apollonius, one of the greatest geometers of antiquity. He conceived that, in the circumference of a circle, having the earth for its centre, there moves the centre of a smaller circle, in the circumference of which the planet revolves. The circle surrounding the earth was called the deferent, while the smaller circle, whose centre was always in the circumference of the deferent, was called the epicycle. It is obvious that the motion of a body from west to east, in this small circle, would be alternately direct, stationary, and retrograde, as was explained in a previous chapter to be actually the case with the apparent motions of the planets. The hypothesis, however, is inconsistent with the *phases* of Mercury and Venus, which, being between us and the sun, on both sides of the epicycle, would present their dark sides towards us at both conjunctions with the sun ; whereas, at one of the conjunctions, it is known that they exhibit their discs illuminated. It is, moreover, absurd to speak of a geometrical centre, which has no bodily existence, moving round the earth on the circumference of another circle. In addition to these absurdities, the whole Ptolemaic system is encumbered with the following difficulties : First, it is a mere hypothesis, having no evidence in its favour, except that it explains the phenomena. This evidence is insufficient of itself, since it frequently happens that each of two hypotheses, which are directly opposite to each other, will explain all the known phenomena. But the Ptolemaic system does not even do this, as it is inconsistent with the phases of Mercury and Venus, as already observed. Secondly, now that we are acquainted with the distances of the remoter planets, and especially the fixed stars, the swiftness of motion, implied in a daily revolution of the starry firmament around the earth, renders such a motion wholly incredible. Thirdly, the centrifugal force which would be generated in these bodies, especially in the sun, renders it impossible that they can continue to revolve around the earth as a centre. Absurd, however, as the system of Ptolemy was, for many centuries no great philosophic genius appeared to expose its fallacies, and it therefore guided the faith of astronomers of all countries down to the time of Copernicus.

After the age of Ptolemy, the science made little progress. With the decline of Grecian liberty, the arts and sciences declined also ; and the Romans, then masters of the world, were ever more ambitious to gain conquests over man than over matter ; and they accordingly never produced a single great astronomer. During the middle ages, the Arabians were almost the only astronomers, and they cultivated this noble study chiefly as subsidiary to astrology.

At length, in the fifteenth century, Copernicus arose, and after forty years of intense study and meditation, divined the true system of the world. The reader will recollect that the Copernican system maintains,—1. That the *apparent* diurnal motion of the heavenly bodies, from east to west, is owing to the *real* revolution of the earth on its own axis from west to east ; and 2. That the sun is the centre around which the earth and planets all revolve from west to east. It rests on the following

arguments : In the first place, *the earth revolves on its own axis.*
First, because this supposition is vastly more *simple.* Secondly,
it is agreeable to *analogy,* since all the other planets that afford
any means of determining the question, are seen to revolve on
their axes. Thirdly, the *spheroidal figure* of the earth is the figure
of equilibrium, that results from a revolution on its axis.
Fourthly, the *diminished weight* of bodies at the equator indicates
a centrifugal force arising from such a revolution. Fifthly, bodies
let fall from a high eminence, fall *eastward of their base,* indicating
that when further from the centre of the earth they were subject
to a greater velocity, which, in consequence of their inertia, they
do not entirely lose in descending to the lower level.

In the second place, *the planets, including the earth, revolve about
the sun.* First, the *phases* of Mercury and Venus are precisely
such as would result from their circulation around the sun in
orbits within that of the earth ; but they are never seen in
opposition, as they would be, if they circulate around the earth.
Secondly, the superior planets do indeed revolve around the
earth ; but they also revolve around the sun, as is evident from
their phases, and from the known dimensions of their orbits ;
and that the sun, and not the earth, is the *centre* of their motions,
is inferred from the greater symmetry of their motions, as
referred to the sun, than as referred to the earth ; and especially
from the laws of gravitation, which forbid our supposing that
bodies so much larger than the earth, as some of these bodies
are, can circulate permanently around the earth, the latter
remaining all the while at rest.

In the third place, the annual motion of *the earth* itself is
indicated also by the most conclusive arguments. For, first,
since all the planets, with their satellites and the comets, revolve
about the sun, analogy leads us to infer the same respecting the
earth and its satellite, as those of Jupiter and Saturn, and indicates that it is a law of the solar system that the smaller bodies
revolve about the larger. Secondly, on the supposition that the
earth performs an annual revolution around the sun, it is
embraced along with the planets, in Kepler's law, that the
squares of the times are as the cubes of the distances ; otherwise, it forms an exception, and the only known exception, to
this law.

Such are the leading arguments upon which rests the Copernican system of astronomy. They were, however, only very

U

partially known to Copernicus himself, as the state both of mechanical science, and of astronomical observation, was not then sufficiently matured to show him the strength of his own doctrine, since he knew nothing of the telescope, and nothing of the principle of universal gravitation. The evidence of this beautiful system being left by Copernicus in so imperfect a state, and indeed his own reasonings in support of it being tinctured with some errors, we need not so much wonder that Tycho Brahe, who immediately followed Copernicus, did not give it his assent, but, influenced by certain passages of Scripture, he still maintained, with Ptolemy, that the earth is in the centre of the universe ; and he accounted for the diurnal motions in the same manner as Ptolemy had done, namely, by an actual revolution of the whole host of heaven around the earth every twenty-four hours. But he rejected the scheme of deferents and epicycles, and held that the moon revolves about the earth as the centre of her motions ; but that the sun, and not the earth, is the centre of the planetary motions ; and that the sun, accompanied by the planets, moves around the earth once a year, somewhat in the manner in which we now conceive of Jupiter and his satellites as revolving around the sun. This system is liable to most of the objections that lie against the Ptolemaic system, with the disadvantage of being more complex.

Kepler and Galileo, however, as appeared in the sketch of their lives, embraced the theory of Copernicus with great avidity, and all their labours contributed to swell the evidence of its truth. When we see with what immense labour and diffi-culty the disciples of Ptolemy sought to reconcile every new phenomenon of the heavens with their system, and then see how easily and naturally all the successive discoveries of Galileo and Kepler fall in with the theory of Copernicus, we feel the force of the lines of the poet :

> " O how unlike the complex works of man,
> Heaven's easy, artless, unincumbered plan."
>
> COWPER.

Newton received the torch of truth from Galileo, and trans-mitted it to his successors, with its light enlarged and purified ; and since that period, every new discovery, whether the fruit of refined instrumental observation or of profound mathematical analysis, has only added lustre to the glory of Copernicus.

With Newton commenced a new and wonderful era in astronomy, distinguished above all others, not merely for the production of the greatest of men, but also for the establishment of those most important auxiliaries to our science, the Royal Society of London, the Academy of Sciences at Paris, and the Observatory of Greenwich. I may add the commencement of the Transactions of the Royal Society, and the Memoirs of the Academy of Sciences, which have been continued to the present time,—both precious storehouses of astronomical riches. The Observatory of Greenwich, moreover, has been under the direction of an extraordinary succession of great astronomers. Their names are Flamstead, Halley, Bradley, Maskeleyne, Pond, and Airy,—the last being still at his post, and worthy of continuing a line so truly illustrious. The observations accumulated at this celebrated Observatory are so numerous, and so much superior to those of any other institution in the world, that it has been said that astronomy would suffer little, if all other contemporary observations of the same kind were annihilated. Sir William Herschel, however, laboured chiefly in a different sphere. The Astronomers Royal devoted themselves not so much to the discovery of new objects among the heavenly bodies, as to the exact determination of the places of the bodies already known, and to the development of new laws or facts among the celestial motions. But Herschel, having constructed telescopes of far greater reach than any ever used before, employed them to sound new and untried depths in the profundities of space. We have already seen what interesting and amazing discoveries he made of double stars, clusters, and nebulæ.

The English have done most for astronomy in observation and discovery; but the French and Germans, in developing, by the most profound mathematical investigation, the great laws of physical astronomy.

It only remains to inquire, whether the Copernican system is now to be regarded as a full exposition of the "Mechanism of the Heavens," or whether there subsist higher orders of relations between the fixed stars themselves.

The revolutions of the *binary stars* afford conclusive evidence of at least subordinate systems of suns, governed by the same laws as those which regulate the motions of the solar system. The *nebulæ* also compose peculiar systems, to which the members are evidently bound together by some common relation.

In these marks of organisation,—of stars associated together in clusters; of sun revolving around sun; and of nebulæ disposed in regular figures,—we recognise different members of some grand system, links in one great chain that binds together all parts of the universe; as we see Jupiter and his satellites combined in one subordinate system, and Saturn and his satellites in another,—each a vast kingdom, and both uniting with a number of other individual parts, to compose an empire still more vast.

This fact being now established, that the stars are immense bodies, like the sun, and that they are subject to the laws of gravitation, we cannot conceive how they can be preserved from falling into final disorder and ruin, unless they move in harmonious concert like the members of the solar system. Otherwise, those that are situated on the confines of creation, being retained by no forces from without, while they are subject to the attraction of all the bodies within, must leave their stations and move inward with accelerated velocity; and thus all the bodies in the universe would at length fall together in the common centre of gravity. The immense distance at which the stars are placed from each other would indeed delay such a catastrophe; but this must be the ultimate tendency of the material world, unless sustained in one harmonious system by nicely-adjusted motions. To leave entirely out of view our confidence in the wisdom and preserving goodness of the Creator, and reasoning merely from what we know of the stability of the solar system, we should be justified in inferring that other worlds are not subject to forces which operate only to hasten their decay, and to involve them in final ruin.

We conclude, therefore, that the material universe is one great system; that the combination of planets with their satellites constitutes the first or lowest order of worlds; that next to these, planets are linked to suns; that these are bound to other suns, composing a still higher order in the scale of being; and finally, that all the different systems of worlds move around their common centre of gravity.

I intended, before concluding this popular digest of astronomical science, to discuss the arguments which astronomy affords to natural theology; but these chapters have been already extended so much further than I anticipated, that I shall conclude with suggesting a few of those moral and religious reflections, which ought always to follow in the train of such a survey of the heavenly bodies as we have now taken.

Although there is evidence enough in the structure, arrangement, and laws, which prevail among the heavenly bodies, to prove the *existence* of God, yet I think there are many subordinate parts of His works far better adapted to this purpose than these, being more fully within our comprehension. It was intended, no doubt, that the evidence of His being should be accessible to all His creatures, and should not depend on a kind of knowledge possessed by comparatively few. The mechanism of the eye is probably not more perfect than that of the universe; but we can analyse it better, and more fully understand the design of each part. But the existence of God being once proved, and it being admitted that he is the Creator and Governor of the world, then the discoveries of astronomy are admirably adapted to perform just that office in relation to the Great First Cause, which is assigned to them in the Bible, namely, "to declare the glory of God, and to show His handiwork." In other words, the discoveries of astronomy are peculiarly fitted—more so, perhaps, than any other department of creation—to exhibit the unity, power, and wisdom, of the Creator.

The most modern discoveries have multiplied the proofs of the *unity* of God. It has usually been offered as sufficient evidence of the truth of this doctrine, that the laws of nature are found to be uniform when applied to the utmost bounds of the *solar system;* that the law of gravitation controls alike the motions of Mercury and those of Neptune; and that its operation is one and the same upon the moon, and upon the satellites of Saturn. It was, however, impossible until recently to predicate the same uniformity in the great laws of the universe respecting the starry worlds, except by a feeble analogy. However improbable, it was still possible, that in these distant worlds other laws might prevail, and other lords exercise dominion. But the discovery of the revolutions of the binary stars, in exact accordance with the law of gravitation, not merely in a single instance, but in many instances; in all cases, indeed, wherever those revolutions have advanced so far as to determine their law of action, gives us demonstration, instead of analogy, of the prevalence of the same law among the other systems as that which rules in ours.

The marks of a still higher organisation in the structure of

clusters and nebulæ, all bearing that same characteristic union
of resemblance and variety which belongs to all the other works
of creation that fall under our notice, speak loudly of one, and
only one, grand design. Every new discovery of the telescope,
therefore, has added new proofs to the great truth that God is
one : nor, so far as I know, has a single fact appeared that is
not entirely consonant with it. Light, moreover, which brings us
intelligence, and, in most cases, the only intelligence we have of
these remote orbs, testifies to the same truth, being similar in
its properties, and uniform in its motions, from whatever star it
emanates.

In displays of the *power* of Jehovah, nothing can compare
with the starry heavens. The magnitudes, distances, and velocities of the heavenly bodies are so much beyond every thing of
this kind which belongs to things around us, from which we
borrowed our first ideas of these qualities, that we can scarcely
avoid looking with incredulity at the numerical results to which
the unerring principles of mathematics have conducted us.
And when we attempt to apply our measures to the fixed stars,
and especially to the nebulæ, the result is absolutely overwhelming : the mind refuses its aid in our attempts to grasp the
great ideas. Nor less conspicuous, among the phenomena of
the heavenly bodies, is the *wisdom* of the Creator. In the first
place, this attribute is everywhere exhibited *in the happy adaptation of means to their ends.* No principle can be imagined more
simple, and at the same time more effectual to answer the
purposes which it serves, than gravitation. No position can be
given to the sun and planets so fitted, as far as we can judge, to
fulfil their mutual relations, as that which the Creator has given
them. I say as far as we can judge ; for we find this to be the
case in respect to our own planet and its attendant satellite, and
hence have reason to infer that the same is the case in the other
planets, evidently holding, as they do, a similar relation to the
sun. Thus the position of the earth at just such a distance from
the sun as suits the nature of its animal and vegetable kingdoms,
and confining the range of solar heat, vast as it might easily
become, within such narrow bounds ; the inclination of the
earth's axis to the plane of its orbits, so as to produce the
agreeable vicissitudes of the seasons, and increase the varieties
of animal and vegetable life, still confining the degree of inclination so exactly within the bounds of safety, that, were it much

to transcend its present limits, the changes of temperature of the different seasons would be too sudden and violent for the existence of either animals or vegetables; the revolution of the earth on its axis, so happily dividing time into hours of business and of repose; the adaptation of the moon to the earth, so as to afford to us her greatest amount of light just at the times when it is needed most, and giving to the moon just such a quantity of matter, and placing her at just such a distance from the earth, as serves to raise a tide productive of every conceivable advantage, without the evils which would result from a stagnation of the waters on the one hand, or from their overflow on the other;—these are a few examples of the wisdom displayed in the mutual relations instituted between the sun, the earth, and the moon.

In the second place, similar marks of wisdom are exhibited in *the many useful and important purposes which the same thing is made to serve.* Thus the sun is at once the great regulator of the planetary motions, and the fountain of light and heat. The moon both gives light by night, and raises the tides. Or, if we would follow out this principle where its operations are more within our comprehension, we may instance the *atmosphere.* When man constructs an instrument, he deems it sufficient if it fulfils one single purpose; as the watch, to tell the hour of the day, or the telescope, to enable him to see distant objects; and had a being like ourselves made the atmosphere, he would have thought it enough to have created a medium so essential to animal life, that to live is to breathe, and to cease to breathe is to die. But beside this, the atmosphere has manifold uses, each entirely distinct from all the others. It conveys to plants, as well as animals, their nourishment and life; it tempers the heat of summer with its breezes; it binds down all fluids, and prevents their passing into the state of vapour; it supports the clouds, distils the dew, and waters the earth with showers; it multiplies the light of the sun, and diffuses it over earth and sky; it feeds our fires, turns our machines, wafts our ships, and conveys to the ear all the sentiments of language, and all the melodies of music.

In the third place, the wisdom of the Creator is strikingly manifested in the provision he has made for the *stability of the universe.* The perturbations occasioned by the motions of the planets, from their action on each other, are very numerous, since every body in the system exerts an attraction on every other, in conformity with the law of universal gravitation. Venus and Mercury, approaching, as they do at times, comparatively near to the earth, sensibly disturb its motions; and the satellites of the remoter planets greatly disturb each other's movements. Nor was it possible to endow this principle with the properties it has, and make it operate as it does in regulating the motions of the world, without involving such an incident. On this subject, Professor Whewell, in his excellent work composing one of the Bridgewater Treatises, remarks: "The derangement which the planets produce in the motion of one of their number will be very small, in the course of one revolution; but this gives us no security that the derangement may not become very large, in the course of many revolutions. The cause acts perpetually, and it has the whole extent of time to work in. Is it not easily conceivable, then, that, in the lapse of ages, the derangements of the motions of the planets may accumulate, the orbits may change their form, and their mutual distances may be much increased or diminished? Is it not possible that these changes may go on without limit, and end in the complete subversion and ruin of the system? If, for instance, the result of this mutual gravitation should be to increase considerably the eccentricity of the earth's orbit, or to make the moon approach continually nearer and nearer to the earth, at every revolution, it is easy to see that, in the one case, our year would change its character, producing a far greater irregularity in the distribution of the solar heat; in the other, our satellite must fall to the earth, occasioning a dreadful catastrophe. If the positions of the planetary orbits, with respect to that of the earth, were to change much, the planets might sometimes come very near us, and thus increase the effect of their attraction beyond calculable limits. Under such circumstances, 'we might have years of unequal length, and seasons of capricious temperature; planets and moons, of portentous size and aspect, glaring and disappearing at uncertain intervals; tides, like deluges, sweeping over whole continents; and perhaps the collision of two of the planets, and the consequent destruction of all organisation on both of them.' The fact really is, that changes are taking place in the motions of the heavenly bodies, which have gone on progressively, from the first dawn of science. The eccentricity of the earth's orbit has been diminishing from the earliest observations to our times.

The moon has been moving quicker from the time of the first recorded eclipses, and is now in advance, by about four times her own breadth, of what her own place would have been if it had not been affected by this acceleration. The obliquity of the ecliptic also is in a state of diminution, and is now about two fifths of a degree less than it was in the time of Aristotle."

But, amid so many seeming causes of irregularity and ruin, it is well worthy of note, that effectual provision is made for the *stability of the solar system.* The full confirmation of this fact is among the grand results of physical astronomy. Newton did not undertake to demonstrate either the stability or the instability of the system. The decision of this point required a great number of preparatory steps and simplifications, and such progress in the invention and improvement of mathematical methods, as occupied the best mathematicians of Europe for the greater part of the last century. Towards the end of that time, it was shown by La Grange and La Place that the arrangements of the solar system are stable; that, in the long run, the orbits and motions remain unchanged; and that the changes in the orbits, which take place in shorter periods, never transgress certain very moderate limits. Each orbit undergoes deviations on this side and on that side of its average state; but these deviations are never very great, and it finally recovers from them, so that the average is preserved. The planets produce perpetual perturbations in each others' motions; but these perturbations are not indefinitely progressive, but periodical, reaching a maximum value, and then diminishing. The periods which this restoration requires are, for the most part, enormous,—not less than thousands, and in some instances millions, of years. Indeed, some of these apparent derangements have been going on in the same direction from the creation of the world. But the restoration is, in the sequel, as complete as the derangement; and in the meantime the disturbance never attains a sufficient amount seriously to affect the stability of the system. "I have succeeded in demonstrating," says La Place, "that, whatever be the masses of the planets, in consequence of the fact that they all move in the same direction, in orbits of small eccentricity, and but slightly inclined to each other, their secular irregularities are periodical, and included within narrow limits; so that the planetary system will only oscillate about a mean state, and will never deviate from it, except by a very small quantity. The ellipses of the

planets have been, and always will be, nearly circular. The
ecliptic will never coincide with the equater; and the entire
extent of the variation, in its inclination, cannot exceed three
degrees."

To these observations of La Place, Professor Whewell adds the
following, on the importance, to the stability of the solar system,
of the fact that those planets which have *great masses* have orbits
of *small eccentricity:*—"The planets Mercury and Mars, which
have much the largest eccentricity among the old planets, are
those of which the masses are much the smallest. The mass of
Jupiter is more than two thousand times that of either of these
planets. If the orbit of Jupiter were as eccentric as that of
Mercury, all the security for the stability of the system, which
analysis has yet pointed out, would disappear. The earth and
the smaller planets might, by the near approach of Jupiter at his
perihelion, change their nearly circular orbits into very long
ellipses, and thus might fall into the sun, or fly off into remoter
space. It is further remarkable, that in the newly-discovered
planets, of which the orbits are still more eccentric than that of
Mercury, the masses are still smaller, so that the same provision
is established in this case also."

With this hasty glance at the unity, power, and wisdom of the
Creator, as manifested in the greatest of His works, I close. I
hope enough has been said to vindicate the sentiment that called
"Devotion, daughter of Astronomy!" I do not pretend that
this, or any other science, is adequate of itself to purify the
heart, or to raise it to its Maker; but I fully believe that, when
the heart is already under the power of religion, there is something in the frequent and habitual contemplation of the heavenly
bodies, under all the lights of modern astronomy, very favourable
to devotional feelings, inspiring, as it does, humility, in unison
with an exalted sentiment of grateful adoration.

THE CRAB NEBULA.

As seen through Lord Rosse's Telescope.

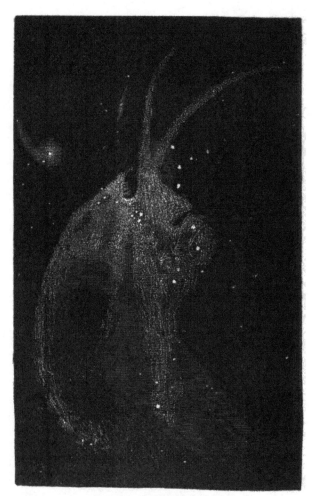

THE GREAT NEBULA IN ORION.

As seen through Lord Rosse's Telescope.

THE NEBULÆ.

(DESCRIPTION OF THE PLATES.)

PLATE I.—LORD ROSSE'S CRAB NEBULA.—This object had long been known as a faint, ill-defined nebula of an elliptical figure, with nothing to distinguish it from many other similar luminous clouds. When examined by Lord Rosse's three-feet reflector, its aspect was entirely changed. The central part was seen to be composed of a vast multitude of stars, while branching filaments of hazy light extended from the centre in all directions. There can be little doubt of the actual character of this singular object. It is probably an immense cluster of suns, of a figure which seems to bid defiance to all our ideas of regularity or order. As it still rests on a bed of misty light, indicating still more distant portions of it, a higher power may hereafter work a complete change in the character of its outline. It is among the most distant objects visible in the telescope, and exhibits a specimen of the singular variety of form, occasionally shown among these vast congeries of stars.

PLATE II.—THE GREAT NEBULA IN ORION.—Discovered by Huygens in 1656. It is one of the largest among the nebulæ, and certainly one of the most magnificent. It was examined by the elder Herschel, with the entire range of his telescopic power, without exhibiting any appearance of being composed of stars. Within the outline of the nebula, there are a great many large and brilliant stars, but these are undoubtedly between the eye and the nebula, and in no way connected with it. Among these stars there is one marked on the maps as θ Orionis, which has long been known to consist of four component stars. Professor Struve, of Dorpat, detected a fifth.; and a sixth has recently been added to this remarkable set. Here we have an example of six suns, probably in physical union, and revolving about their common centre of gravity. Near these six stars, and between the short point of light and the long projecting branch, there appears to be an absolute vacuity, intensely *black*. Whether this be an effect of contrast with the brilliancy of the nebula, or occasioned by some peculiar constitution of this region, it is impossible to determine. In exhibiting this object to persons who had never seen anything of the kind, I have frequently heard them remark, that a part of the nebula was hid by a *black cloud*.

Sir W. Herschel being unable to resolve this nebula, with his most powerful telescopes, placed it among those which he regarded as probably composed of the nebulous fluid, or chaotic matter. The subsequent investigations of Sir John Herschel confirmed the views of his father, and it has only been on the application of the six-feet reflector of Lord Rosse, that the true character of this object has been ascertained. The space-penetrating power of the monster telescope has carried the observer near enough to discern the individual stars of which it is composed. Its resolution has also been accomplished by the magnificent Refractor of the Cambridge (U. S.) Observatory.

The distance and magnitude of this object, as thus determined, absolutely overwhelm the mind. In case light be not absorbed in its journey through the celestial spaces, the light of the nebula in Orion cannot reach the eye in less than sixty thousand years, with a velocity of twelve millions of miles in every minute of time! and yet this object may be seen from this stupendous distance even by the naked eye! What, then, must be its dimensions! Here, indeed, we behold a *universe* of itself too vast for the imagination to grasp, and yet so remote as to appear a faint spot upon the sky.

The resolution of this nebula has been the signal for the renunciation of Herschel's nebular hypothesis.

PLATE III.—NEBULA IN THE SHIELD OF SOBIESKI.—This is an irresolvable nebula, figured by Sir John Herschel, during his residence at the Cape of Good Hope. Its favourable position, as seen in southern latitudes, enabled Herschel to trace the outline of the nebula much farther than any preceding observer had done. The singular figure of this object seems to suggest some power of attraction operating on the particles of matter, or the

NEBULA IN THE SHIELD OF SOBIESKI.

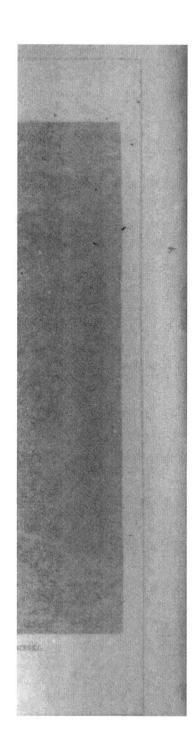

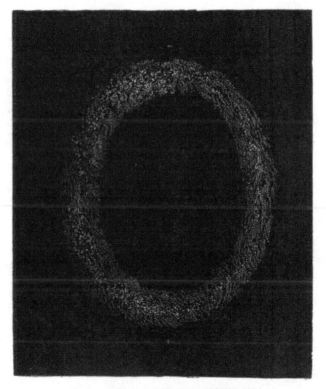

THE RING NEBULA, IN THE CONSTELLATION LYRA.

THE WHIRLPOOL OR SPIRAL NEBULA.
As seen through Lord Rosse's Telescope.

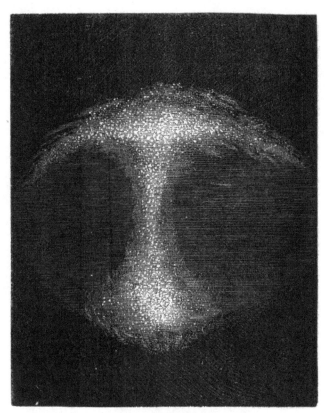

THE DUMB-BELL NEBULA.
As seen through Lord Rosse's Telescope.

stars of which it is composed. In one place we see a bright condensed mass of light, nearly round, while it is encircled by a perfectly dark vacuity, on the outside of which nebulous light is again seen. The smaller curve, at the opposite extremity, was discovered by Herschel, while at the Cape, and has not been seen at any northern observatory, so far as I know. The stars scattered through this nebula are not probably connected with it, but accidental in their position. It is impossible to assert, with certainty, the absolute nebulosity of any object. It may be that higher powers will resolve even this one into stars; but thus far there has been no evidence gained of its resolvable character.

PLATE IV.—THE RING NEBULA, IN THE CONSTELLATION LYRA. —The remarkable size and form of this remote object have combined to render it one of the most curious and interesting in the heavens. It is beautifully shown by the Refractor of the Cincinnati Observatory, and appears as a ring of misty light hung in the heavens, with a diameter as large as that of the moon, when seen with the naked eye. In instruments of inferior power, the interior of this ring appears quite dark; but I have always seen this space filled with more or less light, according to the purity of the atmosphere at the time of examination.

It is supposed to be a vast congeries of stars, united into one grand system. This object is so remote, as to be utterly invisible to the naked eye, and its light cannot reach us in less than twenty or thirty thousand years !

It has shown symptons (as Herschel remarks) of resolvability; but has not been fully resolved. There is one other ring nebula known in the northern heavens, and I believe, but one. Herschel has found several in the south.

PLATE V.—LORD ROSSE'S WHIRLPOOL OR SPIRAL NEBULA.— This is certainly one of the most wonderful objects in the heavens. I have frequently examined it with the Cincinnati Refractor. The principal outlines are well shown, but the filling up is only accomplished by the monster refractor of the Irish nobleman. The two great central clusters are seen with my large instrument, and were originally drawn by Herschel as distinct objects. The curious spiral form is exhibited with great beauty, and seems to indicate the action of some powerful and controlling law, in this remote body or cluster of universes; for such indeed it seems to be.

PLATE VI.—THE DUMB-BELL NEBULA, AS FIGURED BY LORD ROSSE.—This object was discovered by Messier, as early as 1764. It occupies one of the richest portions of the heavens, and is surrounded by thousands of stars. In ordinary telescopes it may be seen in the shape of a double-headed shot, or dumb-bell; hence its name. With more power its outlines are changed, until, with Lord Rosse's great telescope, it is seen under the figure shown in the plate. I have seen the object very nearly under the form in the drawing, but cannot say that I have ever fairly resolved it into stars. The Cambridge Refractor, in the hands of Mr. Bond, is said to have accomplished its resolution. I see many stars on the nebula, but presume them to be accidental, or located between the observer and the nebula.

THE END.

LONDON:
BRADBURY AND EVANS, PRINTERS, WHITEFRIARS.

Printed in the USA
CPSIA information can be obtained
at www.ICGtesting.com
LVHW020205311023
762641LV00011B/242